# TECHNOLOGICAL CHANGE

*Other Works by* Edwin Mansfield

The Economics of Technological Change
Industrial Research and Technological Innovation
Microeconomics: Theory and Applications
Defense, Science, and Public Policy (EDITOR)
Elementary Statistics for Economics and Business (EDITOR)
Managerial Economics and Operations Research (EDITOR)
Monopoly Power and Economic Performance, Revised (EDITOR)

Edwin Mansfield

**WHARTON SCHOOL, UNIVERSITY OF PENNSYLVANIA**

# TECHNOLOGICAL CHANGE

W·W·NORTON & COMPANY·INC·

NEW YORK

SBN 393 09973 3
Library of Congress Catalog Card No. 73-133956

Published simultaneously in Canada
by George J. McLeod Limited, Toronto
PRINTED IN THE UNITED STATES OF AMERICA

1 2 3 4 5 6 7 8 9 0

To Ted,

who is also writing a book

# CONTENTS

# PREFACE

Technological change is of enormous importance at nearly every stage in the study of economics. For example, to understand the fundamentals of economic growth and development, the student must have a basic understanding of technological change. Yet, strange as it may seem, the process of technological change is only recently beginning to receive proper attention in discussions of economic growth. And in elementary economics courses it continues to be neglected, in part because of the lack of proper teaching materials.

The student must also understand the process of technological change in order to appreciate the true nature of competition in many markets. It is obvious that competition often occurs through the introduction of new products and processes. One firm tries to drive another firm out of a particular market by coming out with an improved product that will outperform its competitor's product. In the process of doing this, entire new fields and industries are sometimes created. Yet despite the acknowledged importance of technological change in the birth, death, and competitive struggle of individual firms and industries, this topic receives relatively little attention in elementary textbooks.

Technological change has other ramifications in the study of economics. It is a key factor in the workings of modern labor markets and in questions concerning national policies toward science, defense, and the environment. With regard to labor markets, it is clear that technological change has an impact on the nature and number of jobs in various industries. With regard to military and environmental problems, technological change is of overriding importance. The history of the cold war is, to a considerable extent, the history of enormous efforts at technological advancement in weaponry by the U.S. and the U.S.S.R. As for environmental problems, many of our most pressing difficulties in this area are due to the premature or unwise use of modern technology.

The purpose of this book is to provide a brief overview and interpretation of the economics of technological change. It describes and analyzes the way in which new processes and products are created and assimilated, as well as the public and private policy issues involved. My hope is that it will prove useful in elementary courses. At present, no book exists that covers with the required brevity the range of topics that are relevant for such courses. This book is an abridged version of my *Economics of Technological Change,* but some material (particularly at the end of Chapter 4) has been added, and, where necessary, information has been updated. The present book is designed for courses where it is not feasible or desirable to use the *Economics of Technological Change.* If an instructor can devote sufficient time to the subject, the *Economics of Technological Change* is the more appropriate text, since it deals with the topic in considerably more detail.

*E.M.*

CHAPTER ONE

# The Economics of Technological Change

## 1. Technological Change and the Economy

Without question, technological change is one of the most important determinants of the shape and evolution of the American economy. Technological change has improved working conditions, permitted the reduction of working hours, provided an increased flow of products, old and new, and added many new dimensions to our way of life. The newspapers testify each day to the widespread and profound influence of technological change. Production facilities are automated, educational processes are aided by machines, space vehicles are developed, diseases are conquered, and countless other kinds of changes are made.

Unfortunately, there is also a more somber side to technological change. Advances in military technology have made possible the destruction of mankind on an unprecedented scale, modern technology has resulted in air and water pollution, the closing of plants made obsolete by technological change has thrown whole communities into distress, and the technological revolution in agriculture has contributed to serious problems, both urban and rural. Although most people would agree that, on balance, technological change has been beneficial, no one would claim it has been costless.

This introductory chapter describes briefly some of the principal ways in which the economy is affected by technological change—the advance in knowledge relative to the industrial arts which permits, and is often embodied in, new methods of production, new designs for existing products, and entirely new products and services.[1] We begin by investigating the relationship between technological change and the rate of economic growth. Then we look at the role of technological change in the activities of the Federal government and the effect of technological change on unemployment. Finally, we discuss the importance of technological change in the competition among firms for markets and profits.

## 2. Technological Change and Economic Growth

Technological change is an important, if not the most important, factor responsible for economic growth. The significance of maintaining a high rate of economic growth is widely accepted; target growth rates have been established by the governments of countries with such diverse economies as France, Japan, Sweden, India, Yugoslavia, and the Soviet Union. Even the United States and the United Kingdom, dissatisfied with their past growth records, have set such goals. Economic growth is also an important objective at the international level, the Organization for Economic Cooperation and Development having set as a goal a

[1] A more precise definition of technological change is given in Chapter II, section 1. "Change in technology" and "technological change" are used interchangeably in this book. There is still considerable variation among authors in definitions of terms. For a good discussion, see J. Schmookler, *Invention and Economic Growth,* Cambridge, Mass.: Harvard University Press, 1966, pp. 1–10.

50 percent increase in the collective gross product of the Atlantic Community during the sixties.

Attempts have been made in recent years to measure the effect of a nation's rate of technological change on its rate of economic growth. Several influential studies carried out in the fifties concluded that about 90 percent of the long-term increase in output per capita in the United States was attributable to technological change, increased educational levels, and other factors not directly associated with increases in the quantity of labor and capital. A more recent, and more exhaustive, study concludes that the "advance of knowledge" contributed about 40 percent of the total increase in national income per person employed during 1929–1957. Although these studies are useful, their results are extremely rough. Because of the complex interactions among the various factors that affect the economic development of a country, it is difficult to estimate from historical statistics the precise effect of a nation's rate of technological change on its rate of economic growth. All that can safely be said is that the effect has been substantial.[2]

## 3. Technological Change and the Federal Government

Technological change plays a major role in the activities of the Federal government. Whereas the government's scientific and technical activities were formerly quite small, they now represent

[2] There are at least three problems in these estimates. First, the effects of technological change are measured entirely by the growth of output unexplained by other factors, the consequence being that they are mixed up with the effects of whatever inputs are not included. Second, the use of GNP as a measure of output has a number of important difficulties and misses some of the most important effects of technological change—on leisure and the spectrum of choice. In particular, there are problems in the valuation of entirely new products. Third, these studies fail to recognize the full interdependence of technological change, education, and growth in physical capital with the result that the estimated contribution of each may not be a good indication of the sensitivity of the growth rate to an extra investment in any one of them. See Chapter II, section 9.

a vast enterprise which has important economic and social effects. In the mid-sixties, expenditures for research and development constituted about 15 percent of the Federal administrative budget, much of this research and development being connected with defense. Realizing that any nation which falls significantly behind in military technology will be at the mercy of a more progressive foe, the great powers have spent enormous amounts on military research and development, precipitating several revolutions in technology in the past twenty-five years. Most important have been the successful development and improvement of fission and fusion bombs, although significant achievements have also occurred in delivery vehicles, guidance techniques, radar, and other areas.

The importance of the decisions made regarding military research and development is illustrated by the development of the hydrogen bomb. For a considerable period after World War II, development work on atomic weapons proceeded slowly. However, during the late summer of 1949, the situation changed radically when government scientists found evidence that the Russians had successfully tested an atomic weapon of their own. When the existence of the Russian bomb became known, some members of Congress asked that work be begun to develop a hydrogen fusion bomb, the possibility of which had been discussed extensively at the University of California during World War II. Although eminent scientists advised against a crash program on the ground that there was little point in going beyond the destructive power already available in the A-bomb, a special subcommittee of the National Security Council advised President Truman to inaugurate such a program, which he did early in 1950. From the point of view of national security, this turned out to be a very significant decision. The new bomb was much more powerful than the older one, and only nine months after it had been developed, the Soviet Union produced a similar weapon.

In recent years, many observers have expressed concern over the adequacy of our national science policies. There has been considerable uneasiness regarding the heavy concentration of the nation's scientific resources on military and space work; some, like President Eisenhower, fearing that public policy may become the captive of a scientific elite allied with military and industrial power, others being concerned that other high-priority fields, like transportation and housing, are being deprived of research and development resources. Questions have been raised concerning the efficiency of various government research and development programs, the optimality of the supply of scientific and engineering manpower, the effects of the patent system, and the effectiveness of the Federal decision-making process concerning research and development programs. National science policy is now a matter of widespread interest and considerable concern.

## 4. Technological Change and Unemployment

Changes in techniques can result in the displacement of workers. This is another reason for public interest in technological change, particularly in periods like the late fifties and early sixties when the unemployment rate was relatively high. In 1962, President Kennedy stated that "The major domestic challenge of the sixties is to maintain full employment at a time when automation is replacing men." [3] There was a prominent debate over the extent to which the high employment rates that prevailed then were due to changes in techniques, some economists arguing that technology was advancing more rapidly than in earlier years, that the advances being made were reducing the importance of blue-collar jobs and goods-producing industries relative to white-collar jobs and service-producing industries, and that consequently the unemployed were larger in numbers and out of work longer than in previous years. Other economists believed that the unemployment problem was due largely to an inadequacy of aggregate demand for goods and services. Although neither camp had a monopoly on the truth, statistical studies as well as subsequent events seemed to favor the latter position.

Technological change occupies an important place in the eco-

[3] Quoted in J. Dunlop, *Automation and Technological Change,* The American Assembly, 1962, p. 1.

nomics of labor. The problem of labor displacement is not as prominent now as it was in the early sixties, but it has by no means vanished. Policy makers are still obliged to cope, as best they can, with the changes in the composition and distribution of the labor force induced by technological (and other) change. Collective bargaining is concerned continually with the problem of permitting changes in techniques while protecting worker security. Although unions and companies have experimented with various types of solutions to this problem, it seems likely, in the years immediately ahead, that successful adjustment to technological change will continue to require the best efforts of people on both sides of the bargaining table.

## 5. Technological Change and Industrial Competition

Technological change is a key element in the competitive struggle among firms. The extent and quality of a firm's research and development program can make it an industry leader or head it for bankruptcy. Technological change can transform an industry. For example, a spectacular case in the drug industry was the effect of American Cyanamid's Achromycin tetracycline, introduced in 1953, on sales of Aureomycin chlortetracycline, which had been marketed since late 1948. After an almost continuous upward trend in 1950–1953, Aureomycin sales dropped by nearly 40 percent during the first full year of the sale of Achromycin. This is hardly a typical case, but it illustrates the devastating effect a new product can have on an existing market. In most industries, new products account for a significant share of the market. For instance, in 1960, 10 percent of the sales of all manufacturing firms were accounted for by products developed since 1956.

Recognizing the importance of technological change, firms have increased their outlays on research and development at a rapid rate. This expansion of industrial research and development is one of the most remarkable economic developments of the postwar era. In 1941, industry performed less than $1 billion worth of research and development; in 1953, about $3.5 billion; and in 1967, about $16 billion. Moreover, this is not due entirely to the increased expenditures on research and development by the government; including only company-financed research and development, there has been a tremendous increase in recent years. Turning to the future, most economists seem to believe that spending on research and development will continue to rise, though at a reduced rate.

This "new competition" through research and development has added fuel to the old argument regarding the evils and benefits from giant corporations. Some observers, following the lead of Joseph Schumpeter, have claimed that very large firms are needed to produce the technical achievements on which economic progress depends. For example, according to John Kenneth Galbraith, "technical development has long since become the preserve of the scientist and the engineer. Most of the cheap and simple inventions . . . have been made . . . [Development] can be carried on only by a firm that has the resources associated with considerable size."[4] Needless to say, this proposition has not gone unchallenged by those who feel that technological change (and the rapid acceptance of new techniques) can be achieved without encouraging industrial giantism.

[4] J. K. Galbraith, *American Capitalism*, Boston: Houghton Mifflin Company, 1952, pp. 91–92.

# CHAPTER TWO

# Technological Change and Productivity Growth

## 1. Technological Change, New Techniques, and Scientific Advance

In this chapter, we are concerned with the nature, determinants, and measurement of technological change, as well as with the behavior of various indexes of productivity. We begin by defining more precisely what we mean by technological change. Technology is society's pool of knowledge regarding the industrial arts. It consists of knowledge used by industry regarding the principles of physical and social phenomena (such as the properties of fluids and the laws of motion), knowledge regarding the application of these principles to production (such as the application of genetic theory to breeding of new plants), and knowledge regarding the day-to-day operations of production (such as the rules of thumb of the craftsman). Technological change is the advance of technology, such advance often taking the form of new methods of producing existing products, new designs which enable the production of products with important new characteristics, and new techniques of organization, marketing, and management.

It is important to distinguish between a technological change and a change in technique. A technique is a utilized method of production. Thus, whereas a technological change is an advance in knowledge, a change in technique is an alteration of the character of the equipment, products, and organization which are actually being used. For a technological change to be used, much more is required than the existence of the information. The proper people must possess the information and must be part of an organization which can make effective use of the information. In addition, it is useful to distinguish between technological change and the diffusion of existing information. A new piece of knowledge is a technological change when it is first discovered; but it is not counted as a technological change when it subsequently is passed from one person to another.

It is also important to distinguish between technological change and scientific advance. Pure science is directed toward understanding, whereas technology is directed toward use. Although the distinction between science and technology is imprecise, it should not be ignored. Technological change often occurs as a result of inventions that do not rely on new scientific principles. Indeed, according to historians, little practical use was made of scientific knowledge until the middle of the nineteenth century, when research methods were first used in a systematic way to develop new products in the field of chemistry. The inventions that provided the basis for the industrial revolution were invented by practical men and based upon observation, art, and common sense. The factory, with its machines and its employment of unskilled or semiskilled labor doing simple repetitive operations, and the utilization of materials like iron, coal, and copper are samples of this inventiveness. Turning to the present, it is still true that many changes in technology require no new scientific principles. For example, the zipper and the safety pin required none; neither did the continuous wide strip mill in the steel industry.

Even when changes in technology have been intimately connected with previous scientific breakthroughs, they have not necessarily followed these breakthroughs in any simple and direct way. For example, consider the case of the radio. During the nineteenth century, the developing scientific knowledge of electricity was exploited in a number of areas, the observations of Gilbert, Henry, and Maxwell being seized upon by the inventors of the electric motor, the electric generator, and the telegraph. However, although Maxwell announced his theory of electromagnetism in the 1860's and Hertz's first practical laboratory demonstrations of the production and detection of wireless waves took place in the 1880's, it was not until Marconi formed his company in 1897 that an appreciable amount of applied work began on the radio.

It should be clear, therefore, that changes in technology are quite distinct from scientific advances. Two other points must be added. First, although it is often impossible to connect particular changes in technology with previous scientific advances in any simple way, this does not mean that the character and direction of the advance of science has not affected the rate and direction of technological change. On the contrary, the nature of scientific advance has had a very great influence on the kinds of technological changes we have been able to make. In many important cases, earlier scientific advances have been essential. For example, the discoveries of Faraday, Franklin, and Henry led to the creation of the electrical industry, and those of Chadwick, Fermi, Hahn, Meitner, and Strassmann led to the creation of nuclear technology.

Second, technological change now seems to be more closely related to scientific advance than in the past. The growth of the engineering profession has made a considerable difference in the speed with which new scientific discoveries are translated into changes in technology. (Germany pioneered in engineering education; in the United States engineering schools began to expand rapidly after the Civil War, but they were behind their German counterparts until the twentieth century.) Another factor promoting a closer relationship between science and technology is the growth of industrial and government research laboratories. It is becoming increasingly common for firms and government agencies to sponsor research in fields where improved knowledge is judged likely to open up important areas for development. Much more will be said about the growth of organized research and development in subsequent chapters.

## 2. Technological Change and the Production Function

The technology existing at a given point in time sets limits on how much can be produced with a given amount of inputs. Given the level of technology, there is generally a wide range of possible methods of producing a particular good or service. Some require little capital and much labor, some require much capital and little labor; some are cheap, some are expensive; some old, some new. Moreover, the range is wider than a simple choice among methods that can be taken off the shelf. Other techniques frequently have been explored and could be brought to perfection with only a small amount of development work. Each possible method requires certain inputs—labor, materials, equipment, land—to produce the good or service in question. Given a certain amount of these inputs, it is possible to determine which method results in the maximum output and what the maximum output is.

The production function shows, for a given level of technology, the maximum output rate which can be obtained from given amounts of inputs. For example, if there were only two inputs, capital and labor, Figure 2.1 might show the production function for a particular product at a particular point in time.[1] Each of the curves pertains to a certain level of output, and shows the various combinations of capital and labor that will produce this output. (For example, an output rate of 50 units per year can be achieved by using 20 units of labor and 10 units

[1] Of course, Figure 2.1 shows only part of the production function. There are curves for output levels other than 20, 50, 100, but for simplicity they are omitted from the diagram. An example of a production function is $Y = AL^{a}K^{1-a}$, where $Y$ is the output rate, $L$ is the rate of labor input, and $K$ is the rate of capital input. This is the so-called Cobb-Douglas production function.

of capital per year or by using 16 units of labor and 12 units of capital per year.) Of course, the curve does not show all combinations that can produce a given output. Methods that are technically inefficient—in the sense that, to product the given quantity of output, they use more of one input and at least as much of another input as some other method—are omitted.

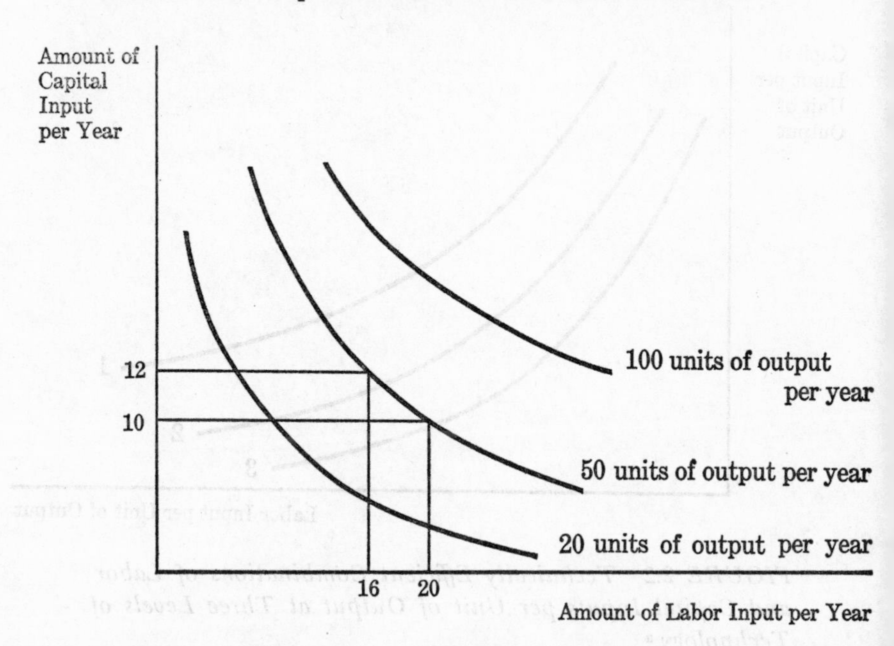


*FIGURE 2.1 Hypothetical Production Function* [a]

[a] See note 1.

Technological change results in a change in the production function. If the production function were readily observable, a comparison of the production function at two points in time would provide the economist with a simple measure of the effect of technological change during the intervening period. If there were constant returns to scale, the characteristics of the production function at a given date could be captured fully by a single curve that would show the various combinations of labor and capital inputs per unit of output that are technically efficient.[2] Under these circumstances, one could simply look at the changing position of this curve. For example, if this curve shifted from position 1 to position 2 in Figure 2.2 during a given period of time, technological change had less impact during this period

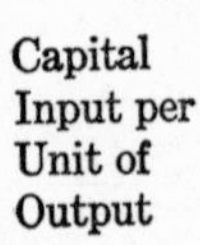


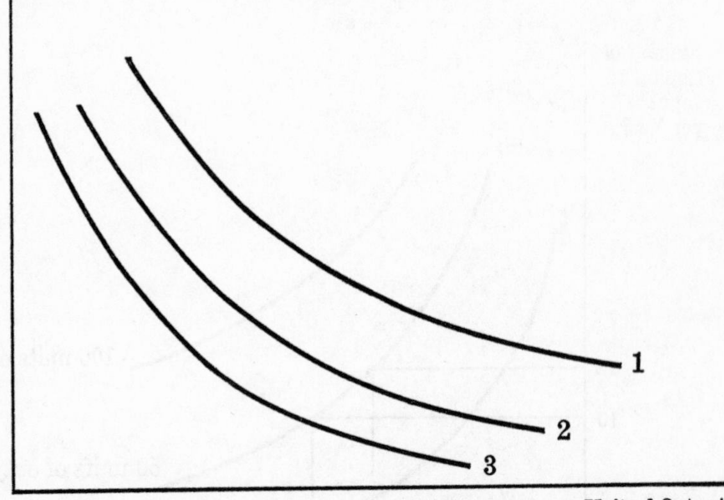


*FIGURE 2.2 Technically Efficient Combinations of Labor and Capital Inputs per Unit of Output at Three Levels of Technology* [a]

[a] Note that this diagram assumes constant returns to scale. See note 2.

than if the curve shifted to position 3. As we shall see in subsequent sections, it is sometimes possible to estimate the average rate of movement of the production function by a single number, and economists often use this number to measure the rate of technological change. Of course, it is only an indirect mea-

2 If there are constant returns to scale, an *x* percent increase in all inputs results in an *x* percent increase in output. Under these conditions, it is clear that the efficient combinations of labor and capital inputs per unit of output fall along a single curve, regardless of the level of output.

sure, but there is no way to measure the rate of technological change directly.[3]

Technological change also results in the availability of new products. In many cases, the availability of new products can be regarded as a change in the production function, since they are merely more efficient ways of meeting old wants, if these wants are defined with proper breadth. This is particularly true in the case of new intermediate goods, which may result in little or no change in the final product. In other cases, however, the availability of new products cannot realistically be viewed as a change in the production function, since they entail an important difference in kind.[4]

The variety of ways in which technological change can influence the production functions in various industries is evident in the development of the electronic computer—a device that incorporates some of the most important technological advances of the twentieth century. Although mechanical tabulating and calculating machines have been in existence for over fifty years, the electronic computer permits information to be processed many times faster and more accurately than any previous method. In the United States, work on the first electronic computer, ENIAC, began at the University of Pennsylvania in 1942 and was completed in 1946. It received financial support from the Army and was mainly designed to calculate trajectories of shells and bombs. After the war, rapid progress was made in solving

[3] The production function shows the best that can be achieved in an industry; it does not show what the average firm—or an inefficient or ill-informed firm—can achieve. Thus, changes in the production function indicate the rate at which the technological frontier for the industry moves forward. It is also important to study the rate at which knowledge is applied and new techniques and products are accepted throughout an industry. These topics are discussed below, particularly in Chapter IV.

[4] It is sometimes asserted that, since firms report that only about 13 percent of their expenditures on research and development go for pure process improvement, it is unrealistic to emphasize cost-reducing technological change. But this is wrong because much of the research and development concerning new products and product improvements is devoted to new and improved intermediate goods and capital goods. In civilian industry, perhaps 80 percent of the reported research and development goes for new processes, new intermediate goods, and new capital goods.

problems of logic design, memory storage systems, and programming techniques. Eventually, the computer became not only technically but economically feasible for a wide range of scientific and business applications.

The development of the computer has resulted in important changes in the production function for many goods and services. For example, in the chemical, petroleum, and steel industries, digital computers are the latest step in the evolution of control techniques. Computers help to determine and enforce the best conditions for process operation, as well as act as data loggers. In addition, they can be programmed to help carry out the complex sequence of operations required to start up or shut down a plant. Use of digital computers for process control began about 1958–59 and has grown rapidly, about 300 systems having been installed or ordered in the United States at the end of 1964. They have resulted in increased production, decreased waste, better control of quality, and reduced chance of damage to equipment.

Banking is another quite different industry where the computer has resulted in an important change in the production function. Two of the most important new devices in banking are reader-sorters—which read and sort documents (particularly checks), sending data via a computer to be recorded on tape—and the computer itself, which processes the information it receives from the reader-sorter. These devices often eliminate the conventional machines and processes for sorting checks, balancing accounts, and computing service charges. They greatly facilitate the handling of checks, high-speed sorters being able to process more than 1500 checks per minute. In addition, electronic bookkeeping machines have taken the place of many of the conventional posting machines; one large bank has reported that its posting errors have decreased two-thirds as a result.

Computers have also resulted in an important change in the production function in scientific research and education. Present machines are ten million times faster than a human being in performing many mathematical calculations and, consequently, make possible scientific studies that formerly were beyond reach. According to current forecasts, computers are likely to become important in other areas in education as well—as tools of research into the learning process, as devices to take over much of the clerical work and information-handling in education, and as ways to increase the student's productivity by permitting individualized instruction. Of course, some areas of instruction (like drill-and-practice systems) are more readily adapted to the computer than others (like tutorial systems), and much more work and experimentation will be required before these forecasts become a reality.

## 3. Determinants of the Rate of Technological Change

What determines the rate of technological change in an industry? Existing theory is still in a relatively primitive state, for it is only recently that economists have begun to give this question the attention it deserves. On a priori grounds, one would expect an industry's rate of technological change to depend to a large extent on the amount of resources devoted by firms, by independent inventors, and by government to the improvement of the industry's technology. The amount of resources devoted by the government depends on how closely the industry is related to the defense, public health, and other social needs for which the government assumes major responsibility; on the extent of the external economies [5] generated by the relevant research and development; and on more purely political factors. The amount of resources devoted by industry and independent inventors depends heavily on the profitability of their use. Econometric studies indicate that the total amount a firm spends on research and development is influenced by the expected profitability of the research and development projects under consideration and that the probability of its accepting a particular research and development project depends on the project's expected returns.

[5] External economies and diseconomies are benefits and costs which accrue to bodies other than the one sponsoring the economic activity in question —which in this case is the firm or agency financing the research and development.

Case studies of particular inventions and studies of patent statistics seem to support this view.

If we accept the proposition that the amount invested by private sources in improving an industry's technology is influenced by the anticipated profitability of the investment, it follows that the rate of technological change in a particular area is influenced by the same kinds of factors that determine the output of any good or service.[6] On the one hand, there are demand factors which influence the rewards from particular kinds of technological change. For example, if a prospective change in technology reduces the cost of a particular product, increases in the demand for the product are likely to increase the returns from effecting this technological change. Similarly, a growing shortage and a rising price of the inputs saved by the technological change are likely to increase the returns from effecting it. As an illustration, consider the history of English textile inventions. During the eighteenth century there was an increase in the demand for yarn, due to decreases in the price of cloth and increased cloth output. This increase in demand, as well as shortages of spinners and increases in their wages, raised the returns to inventions that increased productivity in the spinning processes and directly stimulated the work leading to such major inventions as the water frame, the spinning jenny, and the spinning mule.[7]

On the other hand, there are also supply factors which influence the cost of making particular kinds of technological change. Obviously, whether people try to solve a given problem depends on whether they think it can be solved, and on how costly they think it will be, as well as on the expected payoff if they are successful. The cost of making science-based technological changes

[6] Needless to say, these factors are not the only ones that influence the rate of technological change. As emphasized in subsequent chapters, there is considerable uncertainty in the research and inventive processes, and laboratories, scientists, and inventors are motivated by many factors other than profit. Nonetheless, the factors discussed in this section seem very important.

[7] It is easy to see why an increase in product demand raises the expected returns from an investment in improving the industry's technology. It raises the total, absolute returns from a given percentage cost reduction, and tends to increase the total returns from a given product improvement. For a discussion of the effects of relative factor costs, see section 4.

depends on the number of scientists and engineers in relevant fields and on advances in basic science; for example, advances in physics clearly reduced the cost of effecting changes in technology in the field of atomic energy. In addition, the rate of technological change depends on the amount of effort devoted to making modest improvements that lean heavily on practical experience. Although there is often a tendency to focus attention on the major, spectacular inventions, it is by no means certain that technological change in many industries is due chiefly to these inventions rather than to a succession of minor improvements; for example, Gilfillan has shown that technological change in ship-building has been largely the result of gradual evolution. In industries where this is a dominant source of technological change and where technological change is only loosely connected with scientific advance, one would expect the rate of technological change to depend on the number of people working in the industry and in a position to make improvements of this sort.

Besides being influenced by the quantity of resources an industry devotes to improving its own technology, an industry's rate of technological change depends on the quantity of resources devoted by other industries to the improvement of the capital goods and other inputs it uses. Technological change in an industry that supplies components, materials, and machinery often prompts technological change among its customers. Consider the case of aluminum. For about thirty years after the development of processes to separate aluminum from the ore, aluminum technology remained dormant because of the lack of low-cost electrical power. Technological change in electric power generation, due to Thomas Edison and others, was an important stimulus to the commercial production of aluminum and to further technological change in the aluminum industry. In addition, there is another kind of interdependence among industries. Considerable "spillover" occurs, techniques invented for one industry turning out to be useful for others as well. For example, continuous casting was introduced successfully in the aluminum industry before it was adapted for use in the steel industry. The inventor, Siegfried Junghans, turned his attention to steel after inventing a process for non-ferrous metals, which were easier to cast because of their lower melting points. Similarly, when shell molding was first introduced, its value was thought to be limited to molding non-ferrous items, but recent work indicates that it can be used for ferrous items too.[8]

Other factors which influence an industry's rate of technological change are the industry's market structure, the legal arrangements under which it operates, the attitudes toward technological change of management, workers, and the public, the way in which the firms in the industry organize and manage their research and development, the way in which the scientific and technological activities of relevant government agencies are organized and managed, and the amount and character of the research and development carried out in the universities and in other countries. All of these factors are important—and all will be discussed in subsequent chapters. We describe in sections 7, 8, and 10 the results of several studies that have tried to quantify the effects of some of the factors discussed in this section.[9]

## 4. Labor-Saving and Capital-Saving Technological Change

It is customary for economists to distinguish among neutral, labor-saving, and capital-saving technological change. Suppose that the output rate for a given product, as well as the relative prices of capital and labor, are held constant. If technological change results in a greater percentage reduction in capital input than labor input, it is capital-saving; if it results in a greater percentage reduction in labor input than capital input, it is labor-saving; if it results in an equal percentage reduction in

[8] J. Jewkes, D. Sawers, and R. Stillerman, *The Sources of Invention,* New York: W. W. Norton, revised edition, 1970.

[9] Of course, the factors cited in this and the previous paragraph in the text can be encompassed within the supply-and-demand apparatus discussed above. For a very good description of the factors influencing the rate of technological change, see R. Nelson, M. Peck, and E. Kalachek, *Technology, Economic Growth, and Public Policy,* Washington, D.C.: The Brookings Institution, 1967.

capital and labor inputs, it is neutral. To determine whether or not a technological change is labor-saving, it is not enough to know that labor requirements per unit of output have decreased. For example, although the assembly line principle is sometimes considered labor-saving for this reason, this does not prove the point, because the assembly line principle also reduced capital requirements per unit of output by saving floor space and inventories. The important question is whether labor requirements decreased by a greater percentage than capital requirements.[10]

What determines whether technological change is labor-saving, capital-saving, or neutral? Obviously, a firm that attempts to improve technology in a particular area cannot determine very precisely the kind of technological change, if any, that will result from many of its efforts. But to the extent that the firm can influence the results, what determines whether it aims for labor-saving, capital-saving, or neutral technological change? This question has received considerable attention, but the proposed answers suffer from important limitations. In recent discussions of this subject, it is generally assumed that there is a reasonably well defined set of technological changes that can be obtained from a given research and development budget, that the trade-

[10] Using a diagram like Figure 2.2, one can illustrate the effects of a labor-saving, capital-saving, or neutral technological change on the production function. For example, if the combination of inputs used currently is represented by point *A* and if technological change results in a movement

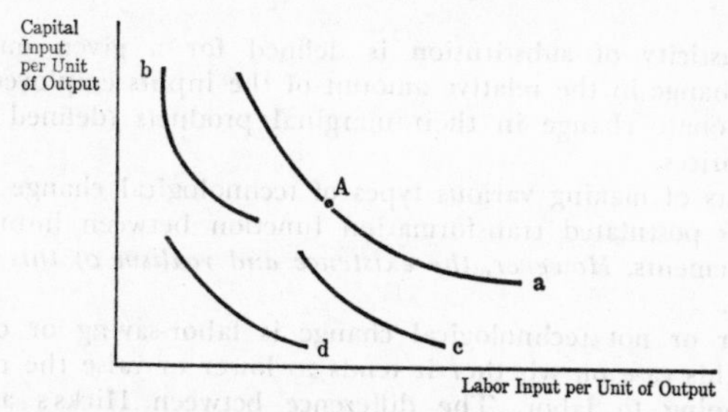


from position a to position b, it is labor-saving; if it results in a movement from position a to position c, it is capital-saving; if it results in a movement from position a to position d, it is neutral. For some relevant discussion, see M. Blaug, "A Summary of the Theory of Process Innovation," *Economica*, February 1963.

offs between various types of technological change are known and constant over time, and that the elasticity of substitution is less than one.[11] Under these circumstances, if an attempt is made to achieve the greatest possible reduction in the total unit cost of production of the product in question, the choice will depend on the percentage of total costs represented by labor costs and on the relative costs of making various types of technological change. In particular, as the ratio of labor costs to total costs increases, the more labor-saving will be the sought-after changes in technology; as the ratio decreases, the more capital-saving they will be. An increase in the cost of making labor-saving changes in technology (relative to the cost of making capital-saving changes in technology) causes the firm to seek more capital-saving changes in technology; a decrease in this relative cost causes it to seek more labor-saving changes.[12]

The United States has experienced for some time an increase in the wage rate relative to the cost of new machinery. According to many economists and economic historians, this should induce labor-saving technological change. This view accords, I suspect, with what most laymen would guess to be the case. However, it is surprising how little evidence there is to support this view. Of course, one complication is that the definition of a labor-saving technological change must be altered when we consider the economy as a whole, since input prices can no longer be taken as given.[13] But this is only part of the problem. (Note that

[11] The elasticity of substitution is defined for a given output as the percentage change in the relative amount of the inputs employed divided by the proportionate change in their marginal products (defined in note 15) or relative prices.

[12] The costs of making various types of technological change are reflected in Kennedy's postulated transformation function between improvements in input requirements. *However, the existence and realism of this function are questionable.*

[13] Whether or not technological change is labor-saving or capital-saving depends in this case on whether it tends to lower or raise the relative share of output going to labor. The difference between Hicks's and Harrod's definition need not concern us here. See J. Hicks, *The Theory of Wages*, London, 1932; J. Robinson, "The Classification of Inventions," *Review of Economic Studies*, 1937–1938; and R. Harrod, *Towards a Dynamic Economics*, London, 1948. Both in this and the preceding paragraphs in the text, we are hampered by the unsatisfactory state of existing knowledge regarding the elasticity of substitution.

it is not possible to conclude that technological change is labor-saving because new techniques are intricate and round-about or because big, heavy equipment is used.) Many economists have the feeling that technological change has been quite labor-saving, but they generally acknowledge that the evidence is indirect and too weak to permit a clear-cut judgment.

## 5. Productivity Growth

Since the days of Adam Smith, economists and policy-makers have been interested in productivity—the ratio of output to input. The oldest and most commonly studied measure of productivity is labor productivity, that is, output per man-hour of labor.[14] Obviously, changes in labor productivity are of importance, since they are intimately related to, though by no means synonymous with, changes in a nation's standard of living. One determinant of the rate of growth of labor productivity is the rate of technological change. In a particular industry or in the entire economy, a rapid rate of technological change is likely to result, all other things being equal, in a high rate of growth of labor productivity. However, since the rate of technological change is not the only determinant of the rate of growth of labor productivity, the latter is a very incomplete, though frequently used, measure of the rate of technological change.

Another important factor influencing the rate of growth of labor productivity is the extent to which capital is substituted for labor in response to changes in relative input prices; obviously, increases in the amount of capital per worker will increase labor productivity. Also, increases in labor productivity may arise because of economies of scale or increases in the extent to which productive capacity is used. In addition, since there is

[14] Needless to say, there is no necessary implication in the term "labor productivity" that all, or any particular part of it, stems from greater effort on the part of workers. This should be obvious from the ensuing discussion in the text. The empirical results in this section are based largely on J. Kendrick, *Productivity Trends in the United States,* Princeton, N.J: Princeton University Press, 1961, which is a very valuable and comprehensive study of productivity statistics in this country.

often a considerable gap between labor productivity with best-practice techniques and labor productivity with the existing mix of techniques, the rate of growth of labor productivity depends on the rate of diffusion of the best practices. Finally, the rate of growth of labor productivity depends on the nature, as well as on the rate, of technological change, labor-saving technological change resulting in greater increases in labor productivity than capital-saving or neutral technological change.

Despite its inadequacies as a measure of the rate of technological change, it is worthwhile giving a brief account of the behavior of labor productivity in the United States. During 1889–1957, the nation's real output per man-hour increased at an average rate of about 2 percent per year. The productivity gains were widely diffused, real hourly earnings growing about as rapidly, on the average, as output per man-hour. These gains were also used to promote increased leisure, working hours being cut by 20 or 30 percent, on the average, since the turn of the century. In the private economy, output per man-hour grew at an average rate of about 2.4 percent, which is somewhat higher than the rate of growth in the economy as a whole. The relatively low rate of growth of productivity in government may be due to the fact that our measures of government output are extremely poor. For this reason, most economists have more faith in the figures for the private economy than for the economy as a whole.

After World War I there was an increase in the rate of increase of output per man-hour. During 1889–1919 output per man-hour rose at an average rate of 1.6 percent per year; during 1920–1957, it grew at an average rate of 2.3 percent per year. The reason for this increase are by no means clear. It may have been due to the spread of the scientific management movement, the expansion of college and graduate work in business administration, the spread of organized research and development, and the change in immigration policy. There has also been a tendency for output per man-hour to rise more rapidly during some phases of the business cycle than during others. Average year-to-year increases in labor productivity were greater when business was expanding (2.4 percent per year) than when it was contracting (1.3 percent per year); the rate of increase was poorest in the first phases of contraction and highest toward the end of the contraction and the beginning of the expansion. In part, these cyclical changes reflect indivisibilities which cause output to decline more than employment even though employees may be fully occupied. In part, such changes reflect the fact that because of their investment in the employees, firms, expecting that business will improve, retain some of the work force even though they are not fully employed.

There were considerable differences among industries in the rate of increase of output per man-hour, as shown in Table 2.1.

*TABLE 2.1 Average Annual Rates of Change of Output per Unit of Labor Input, Various Sectors of the U.S. Private Domestic Economy, 1899–1953*

| *Sector* | *Estimate (percent)* | *Sector* | *Estimate (percent)* |
|---|---|---|---|
| Farming | 1.7 | Manufacturing | 2.2 |
| Mining | 2.5 | Foods | 1.8 |
| Metals | 2.6 | Beverages | 1.6 |
| Anthracite coal | 0.7 | Tobacco | 5.1 |
| Bituminous coal | 1.7 | Textiles | 2.5 |
| Oil and gas | 3.4 | Apparel | 1.9 |
| Nonmetals | 2.9 | Lumber | 1.2 |
| Transportation | 3.4 | Furniture | 1.3 |
| Railroads | 2.8 | Paper | 2.6 |
| Local transit | 2.4 | Printing | 2.7 |
| Residual transport | 4.1 | Chemicals | 3.5 |
| Communications and |  | Petroleum | 3.8 |
| public utilities | 3.8 | Rubber | 4.3 |
| Telephone | 2.0 | Leather | 1.3 |
| Telegraph | 1.6 | Glass | 2.7 |
| Electric utilities | 6.2 | Primary metals | 2.3 |
| Manufactured gas | 4.7 | Fabricated metals | 2.7 |
| Natural gas | 3.0 | Machinery, non- |  |
| Residual sector | 1.4 | electric | 1.8 |
|  |  | Machinery, electric | 2.4 |
|  |  | Transportation |  |
|  |  | equipment | 3.7 |

Source: *J. Kendrick*, op. cit.

However, it is important to note that these figures are less reliable than the national figures—due partly to errors that tend to cancel out in the more aggregative measures and partly to the fact that the output figures for individual industries are gross, not net, of supplies from other industries. Thus, changes over time in the extent to which an industry manufactures its own supplies can influence the labor productivity index. Finally, a relatively great increase in labor productivity in an industry generally meant lower relative costs, lower relative prices, and a better-than-average increase in the volume of production. Better-than-average increases in the volume of production were generally accompanied by better-than-average increases in the level of employment, despite the relatively great increase in output per man-hour. Correspondingly, relatively low increases in labor productivity were usually accompanied by less-than-average increases in output and employment.

## 6. Interplant Differences in Productivity and the Coexistence of Old and New Techniques

At a given point in time, there often are large differences in labor productivity among plants in a particular industry. For example, Table 2.2 shows that, during the first quarter of this century, output per man-hour in blast furnaces using the most up-to-date techniques was generally at least twice as large as the industry average. In addition, the difference between best-practice and average labor productivity often varies considerably among the operations included in an industry. For example, Table 2.3 shows that in the postwar cotton yarn and cloth industry the difference was small for some operations (like card tending and loom fixing) and large for others (like spinning and weaving).

In part, the interplant differences in productivity are due to the coexistence of a variety of techniques in a particular in-

*TABLE 2.2 Best-Practice and Average Labor Productivity, U.S. Blast Furnace Industry, 1911–1926*

| | Gross Tons of Pig-Iron Produced per Man-Hour | |
|---|---|---|
| *Year* | *Best-Practice Plants* | *Industry Average* |
| 1911 | 0.313 | 0.140 |
| 1917 | 0.326 | 0.150 |
| 1919 | 0.328 | 0.140 |
| 1921 | 0.428 | 0.178 |
| 1923 | 0.462 | 0.213 |
| 1925 | 0.512 | 0.285 |
| 1926 | 0.573 | 0.296 |

Source: *U.S. Bureau of Labor Statistics,* The Productivity of Labor in Merchant Blast Furnaces, *1928.*

*TABLE 2.3 Best-Practice and Average Labor Productivity, U.S. Cotton Yarn and Cloth Industry, 1946*

| | Pounds of Cotton Processed per Man-Hour | |
|---|---|---|
| *Operation* | *Best-Practice Plants* | *Industry Average* |
| Picking | 985 | 575 |
| Card tending | 296 | 272 |
| Drawing frame | 493 | 461 |
| Spinning | 86 | 53 |
| Doffing | 141 | 115 |
| Slashing | 979 | 545 |
| Weaving | 89 | 56 |
| Loom fixing | 151 | 143 |

Source: *A. Grosse, "The Technological Structure of the Cotton Industry," in W. Leontief and others,* Studies in the Structure of the American Economy, *Oxford University Press, 1953.*

dustry. Petroleum refining provides one illustration. The first commercial process for cracking heavy petroleum fractions to yield gasoline was introduced in 1913 by William Burton of Standard Oil (Indiana). The Burton process was supplanted by the Dubbs and Tube-and-Tank processes, both introduced in the early twenties. Yet for almost a decade after their introduction, the Burton process accounted for at least 10 percent of the total U.S. output of cracked gasoline. The Dubbs, Tube-and-Tank, and other thermal processes were supplanted in turn by the Houdry, Fluid, and other catalytic processes. Yet for about fifteen years after the introduction of the first catalytic processes, the thermal processes continued to account for most of the industry's capacity.

The mixture of techniques in use at any point in time is the result of a complex combination of technological and economic forces. Technological change results in a new product or in a change in the production function—that is, in the list of technically feasible alternative ways to produce an existing product. Whether a change in technology is applied depends on whether, given the prices of the inputs and outputs, it is economically, as well as technically, attractive to producers. Unless its application seems profitable, it will remain only a potentiality, awaiting the day when altered economic circumstances make it profitable. According to Brozen, the "Detroit Automation" of the fifties was a case of this sort. The use of transfer equipment to move work from one automatic machine tool to another and interlocking these tools to get higher utilization, was first considered in 1927 but was not profitable at the time. Only when wage rates had risen and machine tools had become more expensive did it become economically attractive.

Because of technological change, changes in input prices, changes in demand for the product, or other changes, a new method may become profitable for some firms and consequently it may be adopted by them. But this does not mean that all firms in the industry will immediately switch over to the new technique. Because of their investment in existing equipment and because the economic conditions they face may be different from those faced by the users of the new technique, it may not be economical for some firms to adopt the new technique. Moreover, in the absence of strong economic pressures on them, some firms may be slow to appreciate the new opportunities and to adjust to them. In Chapter 4, we shall see that the diffusion of new techniques often takes a considerable period of time.

## 7. Total Productivity Indexes

The total productivity index relates changes in output to changes in both labor and capital inputs, not changes in labor inputs alone. Specifically, this index equals $q/(zl + vk)$, where $q$ is output (as a percent of output in some base period), $l$ is labor input (as a percent of labor input in some base period), $k$ is capital input (as a percent of capital input in some base period), $z$ is labor's share of the value of output in the base period, and $v$ is capital's share of the value of the output in the base period. Substituting values of $q$, $l$, and $k$ over a given period into this formula, one can easily compute the value of the index for that period. As a measure of technological change, this index has important advantages over labor productivity, the most important being that it takes account of the changes over time in the amount of capital inputs. However, it has the disadvantage of assuming that the marginal products [15] of the inputs are altered only by technological change and that their ratios remain constant and independent of the ratios of the quantities of the inputs.

This formula, or variants of it, has been used to estimate the rate of increase of total productivity in the United States for the period 1899–1957, with these results: First, total productivity for the private domestic economy increased by about 1.7 percent per year over the whole period. Second, there seems to have been an increase in the rate of productivity growth to about 2.1 percent per year in the period following World War I. Third, the rate of productivity increase seems to have been higher in com-

[15] Using the notation in note 1, the marginal product of labor is $\delta Y/\delta L$ and the marginal product of capital is $\delta Y/\delta K$.

munications and transportation than in mining, manufacturing, and farming (Table 2.4). Fourth, within manufacturing it seems to have been highest in rubber, transportation equipment, tobacco, chemicals, printing, glass, fabricated metals, textiles, and petroleum (Table 2.4).

*TABLE 2.4 Estimates of Annual Rate of Increase of Total Productivity in Various Sectors of the U.S. Private Domestic Economy, 1899–1953*

| *Sector* | *Estimate (percent per year)* | *Sector* | *Estimate (percent per year)* |
|---|---|---|---|
| Farming | 1.1 | Manufacturing | 2.0 |
| Mining | 2.2 | Foods | 1.7 |
| Metals | 2.2 | Beverages | 1.6 |
| Anthracite coal | 0.7 | Tobacco | 3.5 |
| Bituminous coal | 1.6 | Textiles | 2.4 |
| Oil and gas | 3.0 | Apparel | 1.7 |
| Nonmetals | 2.6 | Lumber | 1.0 |
| Transportation | 3.2 | Furniture | 1.4 |
| Railroads | 2.6 | Paper | 2.3 |
| Local transit | 2.5 | Printing | 2.6 |
| Residual transport | 4.0 | Chemicals | 2.9 |
| Communications and public utilities | 3.6 | Petroleum | 2.4 |
| | | Rubber | 4.1 |
| Telephone | 2.0 | Leather | 1.2 |
| Telegraph | 1.8 | Glass | 2.6 |
| Electrical utilities | 5.5 | Primary metals | 1.9 |
| Manufactured gas | 4.7 | Fabricated metals | 2.6 |
| Natural gas | 2.0 | Machinery, nonelectric | 1.7 |
| Residual sector | 1.3 | Machinery, electric | 2.2 |
| | | Transportation equipment | 3.5 |

Source: *J. Kendrick,* op. cit.

Another study of the rate of productivity advance presents results for the United States, United Kingdom, Germany, Japan, and Canada during the period since World War II.[16] The find-

[16] E. Domar, S. Eddie, B. Herrick, P. Hohenberg, M. Intrilligator, and I. Miyamoto, "Economic Growth and Productivity in the United States, Canada,

*TABLE 2.5 Estimates of Annual Rate of Increase of Total Productivity in United States, Canada, United Kingdom, Germany and Japan*

| Sector | United States 1948–1960 | Canada 1949–1960 | United Kingdom 1949–1959 | Germany 1950–1959 | Japan 1951–1959 |
|---|---|---|---|---|---|
| | (percent per year) | | | | |
| Economy | n.a. | 1.2 | 0.6 | 3.6 | 3.7 |
| Private economy | 1.4 | n.a. | 0.7 | n.a. | 3.8 |
| Private nonfarm economy | n.a. | n.a. | n.a. | n.a. | 3.9 |
| Agriculture | 2.6 | 2.0 | | | |
| Forestry, fishing, | | | 2.0[a] | 4.3 | 1.2 |
| trapping | n.a. | 0.7 | | | |
| Mining, quarrying, wells | n.a. | 0.9 | 0.3 | | −0.6 |
| Manufacturing | 2.6 | 1.4 | 0.7[a] | | 4.1 |
| Construction | n.a. | 0.6 | 0.2[b] | 3.4 | 2.2 |
| Public utilities | | 2.0 | 1.9[c] | | |
| Transportation and | 3.4 | | | | 4.5 |
| communication | | 1.5 | 1.8 | | |
| Wholesale and | | | | 1.5 | |
| retail trade | n.a. | −0.6 | −1.0[b] | | −0.5 |
| Finance, insurance, real estate | n.a. | 0.6 | 0.6 | | 4.1 |
| Other services | | | | 1.4 | |
| Government | n.a. | −0.8 | −2.8 | | 6.7 |

Source: *E. Domar* et al., op. cit.

[a] 1950–1959.
[b] 1953–1959.
[c] 1950–1958.
n.a. Not available.

ings, presented in Table 2.5, indicate that the rate of increase of total productivity was higher in Germany and Japan than in the United States and Canada, and higher in the United States and Canada than in the United Kingdom. However, if capi-

United Kingdom, Germany, and Japan in the Postwar Period," *Review of Economies and Statistics,* February 1964.

tal inputs had been adjusted for under-utilization, the United States, United Kingdom, and Canada might have turned in a better performance. Examination of the results by sector indicates that in Canada and the United Kingdom, agriculture and public utilities had the highest rates of productivity increase; in Germany, agriculture; in Japan and the United States, public utilities, transportation, and communication.

Finally, what factors seem to influence the rate of growth of total productivity in an industry? Apparently an industry's rate of growth of total productivity is related in a statistically significant way to (1) its ratio of research and development expenditures to sales, (2) its rate of change of output level, and (3) the amplitude of its cyclical fluctuation. Specifically, the rate of growth of total productivity increases (on the average) by 0.5 percent for each tenfold increase in the ratio of research and development expenditures to sales and by 1 percent for every 3 percent increase in the industry's growth rate.[17] These empirical results seem to be consistent with the theories in section 3. However, they are somewhat ambiguous; for example, the observed relationship between the rate of productivity growth and the industry's growth rate could be due partly to an effect of the former on the latter. Correlation does not prove causation.

## 8. Other Measures of Technological Change

Economists have tried to devise better measures of the rate of movement of the production function than the total productivity index. These measures rest on somewhat different assumptions about the shape of the production function, the Cobb-Douglas and CES production functions sometimes, but not always, being used.[18] For example, in an important paper published in 1957, Robert Solow provided an estimate of the rate of

[17] N. Terleckyj, *Sources of Productivity Advance,* Ph.D. Thesis, Columbia University, 1960.

[18] The Cobb-Douglas production function is shown in note 1. For a description of the CES production function, see the first reference in note 21.

technological change for the nonfarm economy during 1909–1949.[19] The results suggest that, for the entire period, the average rate of technological change was about 1.5 percent per year. That is, the quantity of output derivable from a fixed amount of inputs increased at about 1.5 percent per year. In addition, Solow found evidence that the average rate of technological change was smaller during 1909–1929 than during 1930–1949. Benton Massell carried out a similar analysis for United States manufacturing, his estimate of the annual rate of technological change during 1919–1955 being about 3 percent.[20] In contrast with Solow, his results show little or no evidence of a higher rate of technological change during the thirties and forties than in previous decades.[21]

The studies by Solow and Massell assume implicitly that technological change is disembodied—that is, that all technological change consists of better methods and organization that improve the efficiency of both old capital and new. Examples of such improvements are various advances in industrial engineering (for example, the development of time and motion studies) and operations research (for example, the development of linear programming). Although technological change of this sort has undoubtedly been of importance, many changes in technology must be embodied in new equipment if they are to be utilized. For example, the introduction of the continuous wide strip mill in

[19] R. Solow, "Technical Change and the Aggregate Production Function," *Review of Economics and Stastistics,* 1957. Solow assumed that there were constant returns to scale, that capital and labor were paid their marginal products, and that technological change was neutral.

[20] B. Massell, "Capital Formation and Technical Change in U.S. Manufacturing," *Review of Economics and Statistics,* May 1960.

[21] Arrow, Chenery, Minhas, and Solow obtained an estimate of the rate of technological change in the non-farm economy, based on Solow's figures for 1909–1949, and somewhat different assumptions about the shape of the curves in Figure 2.2. The result was 1.8 percent, as compared with Solow's estimate of 1.5 percent. See their "Capital-Labor Substitution and Economic Efficiency," *Review of Economics and Statistics,* August 1961.

Another approach to the measurement of technological change entails the comparison of input-output tables. A comparison for 1947–1958 indicates that technological change during that period tended to reduce the differences in input structure distinguishing the major groups of industries. See A. Carter, "The Economics of Technological Change," *Scientific American,* April 1966.

the steel industry and the diesel locomotive in railroads required new investment in plant and equipment. No one has attempted to measure fully the extent to which technological change in recent years has been capital-embodied, as this kind of technological change is called. But the available evidence clearly indicates that a great deal of capital-embodied technological change has taken place.

If technological change is assumed to be capital-embodied, not disembodied, somewhat different methods must be used to estimate the rate of technological change. What do the results look like? Solow has estimated that the rate of technological change in the private economy during 1919–1953 was 2.5 percent per year.[22] This estimate is higher than his earlier estimate based on the assumption that technological change was disembodied. Turning to individual firms, estimates have been provided for ten large chemical and petroleum firms in the postwar period, one set of estimates assuming that technological change was disembodied, the other assuming that it was capital-embodied. At the industry level, estimates for ten manufacturing industries suggest that the rate of capital-embodied technological change during 1946–1962 was highest in motor vehicles and instruments; next highest in food, chemicals, electrical equipment, paper, and apparel; and lowest in machinery, furniture, and glass.[23] Outside manufacturing, the rate of capital-embodied technological change in the railroad industry during 1917–1959 has been estimated at 3 percent per year.[24]

What factors seem to influence the rate of technological change, as measured by the estimated change in the production function? My results, based on data regarding ten large chemical and petroleum firms and ten manufacturing industries in the postwar period, indicate that, both for firms and for industries,

[22] R. Solow, "Investment and Technical Change," *Mathematical Models in the Social Sciences,* ed. by Arrow, Karlin, and Suppes, Stanford, Calif.: Stanford University Press, 1959.

[23] E. Mansfield, *Industrial Research and Technological Innovation,* New York: W. W. Norton, 1968, Chapter IV.

[24] E. Mansfield, "Innovation and Technical Change in the Railroad Industry," *Transportation Economics,* National Bureau of Economic Research, 1965.

the rate of technological change is directly related to the rate of growth of cumulated research and development expenditures made by the firm or industry. If technological change is disembodied, the average effect of a 1 percent increase in the rate of growth of cumulated research and development expenditures is a .1 percent increase in the rate of technological change. If technological change is capital-embodied, it is a .7 percent increase in the rate of technological change. These results are quite compatible with the theories in section 3. Needless to say, however, they are tentative, since they are based on a relatively small amount of data and since, as noted before, correlation does not prove causation.

## 9. Problems of Measurement

It is important that we point out some of the problems in the measures discussed in the previous section. First, these measures of the rate of technological change are indirect measures that look only at the effects of technological change; and since they equate the effects of technological change with whatever increase in output is unexplained by other factors, they do not isolate the effects of technological change alone. In addition, they contain the effects of whatever factors are excluded—which, depending on the particular study, are increases in education, betterment of worker health and nutrition, economies of scale, changes in project mix, or improved allocation of resources. For this reason they are sometimes called "residuals." So far economists have been unable to sort out the effects of "pure" technological change, except perhaps when dealing with individual processes.

Second, there are difficulties in measuring inputs, the measurement of aggregate capital being a particularly nettlesome problem. In addition, it is difficult to adjust for quality changes in inputs; and there are problems in choosing among price deflators.[25] Third, the customary measures often assume that there

[25] For example, see J. Robinson, "Some Problems of Definition and Measurement of Capital," *Oxford Economic Papers*, June 1959. According

are no economies of scale and that technological change is neutral. The restrictiveness of these assumptions is obvious. Fourth, it often is impossible to distinguish capital-embodied from disembodied technological change on the basis of available data. For this and other reasons, there has been some reaction recently against the usefulness of the capital-embodied-technological-change hypothesis, as generally put forth. Fifth, when one compares a number of studies, there is considerable variation in the estimated rates of technological change in particular industries. Apparently the results are quite sensitive to the detailed assumptions that are made and the data that are used.

The moral of this section is clear. Because the available measures are beset by many important problems, they should be used only as very rough guides. We are a long way from having precise measurements of the rate of technological change.

## 10. Patents and Technological Change

The number of patents is sometimes used as a crude index of the rate of technological change, or some important component thereof, in a given field at a certain point in time. Used in this way, patent statistics have important disadvantages. For one thing, the average importance and cost of the patents granted at one time and place may differ from those granted at another time and place. For another, the proportion of the total inventions that are patented may vary considerably. Nonetheless, it is of interest to see what the patent statistics suggest, since they are the basis for some major investigations in this area.

According to a series of studies by Jacob Schmookler, there is a high correlation between the patent rate on capital-goods inventions in an industry and the lagged value of the industry's

to D. Jorgensen and Z. Griliches, there are many measurement errors and errors of aggregation in the measures that are ordinarily used; and these errors inflate the residual. See "The Explanation of Productivity Change," *Review of Economic Studies,* July 1967. One can always "explain" changes in output by changes in input (appropriately measured), but many of these changes in input must themselves be attributed to technological change.

investment or value-added.[26] That is, a high patent rate on capital-goods inventions is associated with a high previous level of investment or value-added. Turning from comparisons over time to comparisons among industries at a given point in time, there seems to be a tendency for the number of patents on capital-goods inventions to be directly related to the level of investment or value-added in an industry. That is, industries with high investment or value-added account for more patents than those with low investment or value-added. Moreover, this relationship persists when the effects of industry size are taken into account.[27]

What are the implications of these findings? In the past, some, though by no means all, economists assumed that the rate of technological change was determined outside the economic system and was independent of economic variables. To the extent that the patent rate is a useful index of the rate of technological change, or some important component thereof, Schmookler's re-

[26] An industry's value-added is its dollar sales minus its purchases of intermediate products from other firms or industries. Griliches and Schmookler found that 84 percent of the variation in the patent rate on process inventions could be explained by the industry's value-added three years before. See their "Inventing and Maximizing," *American Economic Review*, September 1963. For results based on investment data, see J. Schmookler, *Invention and Economic Growth*, Cambridge, Mass.: Harvard University Press, 1966, as well as his paper with Griliches. Also, Schmookler found that a significant correlation exists between patent applications and variable production inputs, both expressed as deviations from trend. Earlier, Graue found considerable correlation, after removing trends from both series, between the level of industrial production and the number of mechanical patents issued. See E. Graue, "Inventions and Production," *Review of Economics and Statistics*, November 1943; and J. Schmookler, "The Level of Inventive Activity," *Review of Economics and Statistics*, May 1954.

[27] J. Schmookler and O. Brownlee, using data for eighteen manufacturing industries, show that this relationship is quite strong, particularly when the patent data are lagged several years behind the value-added data. The coefficient of correlation, which is higher in more recent years, exceeded .9 in 1947. See their "Determinants of Inventive Activity," *American Economic Review*, May 1962. For results based on investment data, see J. Schmookler, *Invention and Economic Growth, op. cit.* However, if a firm's patent rate, including both process and product inventions, is correlated with its sales, the correlation coefficient is only about .6. See F. Scherer, "Firm Size, Market Structure, Opportunity, and the Output of Patented Inventions," *American Economic Review*, December 1965. Also, see Griliches and Schmookler, *op. cit.*

sults seem to contradict this assumption. Going a step further, he concludes that the distribution of inventions according to function, that is, according to the industry expected to use them, is largely determined by demand factors of the sort discussed in section 3. The supply factors, reflecting, for example, advances in basic science, enter in as determinants of the form—mechanical, chemical, electrical, and so on—in which the inventions occur. Put differently, demand conditions determine which industries or consumer activities inventions are made for; supply conditions determine which industries or branches of science and technology inventions are made by. This is an interesting hypothesis, which undoubtedly will be subjected to further tests using other bodies of data.

As an industry grows older, there seems to be a tendency for the rate of patenting to rise first at an increasing rate, then at a decreasing rate, and finally to decline. This pattern has occurred in a wide variety of industries, the ultimate decline in the patent rate being explained in two quite different ways. According to one hypothesis, the technology in any field rather quickly approaches perfection, with the result that fewer important inventions can be made in it and inventors leave the field. According to the other hypothesis, the decrease in the patent rate occurs because of a decrease in the rewards to be gained from technological change in this industry, these rewards being associated with the growth and profitability of the industry. The decline in the patent rate may be due partly to both hypotheses, but the available evidence, though not entirely unambiguous, seems to indicate that the latter hypothesis is more important.[28]

28 See S. Kuznets, *Secular Movements in Production and Prices*, Boston: Houghton Mifflin, 1930; R. Merton, "The Rate of Industrial Invention," *Quarterly Journal of Economics*, May 1935; A. Stafford, *Trends in Invention in Material Culture*, Ph.D. Thesis, University of Chicago, 1950; W. Salter, *Productivity and Technical Change*, Cambridge University, 1960; and J. Schmookler, *ibid.*

# CHAPTER THREE

# Industrial Research and Development

## 1. Science, Technology, and Industrial Laboratories

In the previous chapter, it was pointed out that the rate and direction of technological change depends on the extent and nature of the research and development [1] (R and D) carried out by private industry, government, and universities. This chapter is concerned with industrial research and development. We look at the nature of industrial research and development, the size

[1] According to the National Science Foundation definitions, which are used throughout this book, research and development "includes basic and applied research in the sciences (including medicine) and in engineering, and design and development of prototypes and processes. It does not include quality control, routine product testing, market research, sales promotion, sales service, research in the social sciences or psychology, or other nontechnological activities or technical services." K. Sanow, "Development of Statistics Relating to Research and Development Activities in Private Industry," *Methodology of Statistics on Research and Development,* National Science Foundation, 1959, p. 124. Also see notes 2, 3, and 4 below.

and composition of industry's R and D expenditures, the organization and management of industrial R and D, and the role of various types of firms and independent inventors in the process by which new technology is created.

The first thing to note is that the organized application of science to advance technology is a relatively new thing. Until the middle of the nineteenth century, the connection between science and technology was very loose. The periods when science flourished did not coincide with those when technology was moving ahead most rapidly. When they did flourish together it was not necessarily at the same place. On balance, science was far more indebted to technology than technology was to science. For example, magnetism was known as an empirical fact, and had been used to construct compasses for centuries before the physicists began studying the subject in the eighteenth and nineteenth centuries. During the second half of the last century, science and technology began to draw somewhat closer together and science began to take the lead in some areas. The beginnings of the synthetic drug and dye industries were related to previous discoveries in chemistry. The application of the dynamo was an outgrowth of Faraday's pioneering work that had taken place about forty years before. The telegraph cable was related to scientific work during the first half of the century. Nonetheless, the relation between science and technology during this period was on the whole quite remote.

Toward the end of the nineteenth century, as the connection between science and technology gradually became closer, commercial research laboratories began to appear. The first organized research laboratory in the United States was established by Thomas Edison in 1876. In 1886, Arthur D. Little, an applied scientist and a missionary of applied science to industry, started his firm. Eastman Kodak (1893), B. F. Goodrich (1895), General Electric (1900), and DuPont (1902) were some of the earliest manufacturing firms to establish laboratories; the Bell Telephone System (1907) was among the first utilities to do so. These companies recognized the potential profitability of in-house research activities. For example, in the case of General Electric, the company was impressed by its debt to Faraday, Maxwell, and other European scientists and hoped to establish corresponding lines of inquiry in the United States.

The industrial laboratory constituted a significant departure from previous days when invention was mainly the work of independent inventors like Eli Whitney (cotton gin), Robert Fulton (steamboat), Samuel Morse (telegraph), Charles Goodyear (vulcanization of rubber), and Cyrus McCormick (reaper). These men were responsible for an extremely rich crop of inventions, some of which established whole new industries. As we shall see in subsequent sections, the advent of the industrial laboratory did not result in the total displacement of the independent inventor. On the contrary, independent inventors continue to produce a significant share of the important inventions, although their relative importance seems to have declined.

During the twentieth century, the number of industrial laboratories grew by leaps and bounds. By World War I, there were perhaps 100 industrial research laboratories in the United States, most of them being in the new fields of electricity and chemistry. The number tripled during the war, and with confidence in organized research increased by wartime successes, most of the new laboratories remained in existence during peacetime. The number increased again during World War II, and by 1960 there were more than 5,400 industrial research laboratories.

## 2. The Nature of Industrial Research and Development

Research and development encompasses work of many kinds, and it is important that we identify the various types. First, there is basic research,[2] which is aimed purely at the creation of new

[2] According to the National Science Foundation, basic research includes "research projects which represent original investigation for the advancement of scientific knowledge and which do not have specific commercial objectives, although they may be in fields of present or potential interest to the

knowledge. Its purpose is to permit changes in ways of looking at phenomena and activities, to identify and measure new phenomena, and to create new devices and methods for testing various theories. For example, the biologist who tries to understand how and why certain cells proliferate, without having any particular application in mind, is carrying out basic research. Industrial laboratories carry out some basic research, but it is a very small proportion of their efforts. The principal bastions of basic research in our society are the universities. Second, there is applied research,[3] which is research expected to have a practical pay-off. Projects of this sort might be directed at ways of making steel resist stresses at particular temperatures, ways of inhibiting the growth of streptococci, or ways of obtaining the energy from atomic fission directly as electricity. The distinction between basic and applied research is fuzzy. Essentially, the distinction is based on the motivation of the researcher, basic research being aimed at new knowledge for its own sake, applied research being aimed at practical and commercial advances. In many cases both motives are present, and it is difficult to classify a particular project in this way.

The nature of industrial research projects can be illustrated by the history of the transistor, which was the result of a research project started in 1946 at the Bell Telephone Laboratories. Before and during World War II, Bell and a number of other laboratories supported extensive research on semiconductors. During 1945, William Shockley persuaded the Bell management to intensify its work in this area. A research group composed of physicists, chemists, and metallurgists was formed, the general

reporting company." See K. Sanow, *op. cit.*, p. 124. The definition for government agencies and universities is somewhat different. "Basic research is that type of research which is directed toward increase of knowledge in science. It is research where the primary aim of the investigator is a fuller knowledge or understanding of the subject under study, rather than, as in the case with applied research, a practical application thereof." *Ibid*, p. 75.

3 Applied research includes "research projects which represent investigation directed to discovery of new scientific knowledge and which have specific commercial objectives with respect to either products or processes. Note that this definition of applied research differs from the definition of basic research chiefly in terms of the objective of the reporting company." *Ibid*, p. 124. Note 2 contains the definition for government agencies and universities.

scientific aim of the program being to obtain as complete an understanding as possible of semiconductor phenomena on the basis of atomic theory. Shockley was particularly interested in the prospects for a solid-state amplifier, and experiments were devised to see if such a gadget worked as theory said it should. When it did not, new theories were proposed and new experiments were performed, one of these experiments providing the first indication of the transistor effect.

The transistor is by no means the only research achievement of fundamental importance that has emerged from industrial laboratories. The discovery of electron diffraction and the wave properties of electrons by Davisson and Germer of the Bell Telephone Laboratories brought Davisson the 1937 Nobel Prize in physics. Irving Langmuir of General Electric received the 1932 Nobel Prize in chemistry. Many other achievements could be noted. Nonetheless, it would be a serious mistake to think that most industrial research is concerned with such fundamental and major goals as these projects. On the contrary, much of it is, as one would expect, rather limited in scope and focused closely on particular applications.

Third, there is development,[4] which is aimed at the reduction of research findings to practice. Development projects are of many kinds. The more advanced development projects aim at the construction of entirely new types of products and processes; the more routine development projects, which often constitute the bulk of the total, aim only at relatively minor modification of products already brought into being by previous research and development. By the time a project reaches the development stage, much of the uncertainty regarding its technical feasibility has been removed, but there usually is considerable uncertainty

[4] Development includes "technical activity concerned with non-routine problems which are encountered in translating research findings or other general scientific knowledge into products or processes. It does not include routine technical services to customers or other items excluded from definition of research and development above." Sanow, *op. cit.* The definition for government and universities is "the systematic use of scientific knowledge directed toward the production of useful materials, devices, systems, or methods, including design and development of prototypes and processes. It excludes quality control or routine product testing." *Ibid,* p. 75.

regarding the cost of development, time to completion, and utility of the outcome. The development phase of a project is generally more expensive than the research phase. There is a long road from a preliminary sketch, showing schematically how an invention should work, to the blueprints and specifications for the construction of the productive facilities. The tasks that are carried out depend, of course, on the nature and purpose of the development project. In some cases, various types of experiments must be made, and prototypes must be designed and developed. Frequently, pilot plants are built and the experience with the pilot plant is studied before large-scale production is attempted. The construction of adequate materials and the design of new ways to work with these new materials are sometimes of crucial importance: In the development of the transistor, particularly the silicon transistor, it was important to find materials in a very pure form; and in the development of the jet engine, it was important to find metals which would withstand abnormal stresses and strains.

## 3. Nylon: A Case Study of Industrial Research and Development

A further illustration of the nature of industrial research and development is provided by nylon, one of the most significant inventions in the twentieth century. The story begins in 1927 when DuPont decided to support a program of fundamental research. In accord with the primary aim of this program, which was to discover scientific knowledge regardless of immediate commercial value, the company set aside about $250,000 per year for a number of research projects in chemistry. Wallace H. Carothers, a 32-year-old chemist who had taught at Harvard and Illinois, was offered the direction of the group and joined the firm in 1928. At DuPont, Carothers continued work he had started at Harvard on condensation polymers, and his early work at DuPont yielded considerable fundamental knowledge of polymerization. At first this information was purely of academic value, but something occurred in 1930 during an experiment with a polyester made from ethylene glycol and sebacic acid which was destined to be of great practical value. While cleaning out a reaction vessel in which he had been making one of these polymers, one of Carothers' associates discovered in pulling a stirring rod out of the reaction vessel that he pulled out a fiber. Moreover, he noted its unusual flexibility and strength, as well as its remarkable ability to cold draw. Although fibers of this original superpolymer were not of practical use because they were easily softened by hot water, they suggested that some related compound might possess characteristics suitable for making commercial fibers.

Additional investigation was undertaken. Numerous superpolymers were synthesized, but each was deficient in some important respect. Discouraged, Carothers gave up the project in 1930 and DuPont seriously considered dropping the work. But the director of the Chemical Department urged Carothers to review his findings and continue the work. Carothers turned his attention to the polyamides and found that several polyamides, when extruded through a spinneret improvised from a hypodermic needle, produced filaments from which fibers could be made. These new polyamides seemed extremely promising. After further experimentation, Carothers and his associates developed a polyamide from which strong, tough, elastic, water-resistant fiber could be made. This polymer, which they called "66" polymer, is made by the reaction between hexamethylene diamine and adipic acid. Textile experts chose this polymer for commercial development, and although many other polyamides have been evaluated since, it is still the type most commonly used for textile purposes. The resulting fiber was called nylon.

During the next two years DuPont scientists and engineers tackled "the development on a laboratory scale of the manufacturing processes for the intermediates, the polymer and nylon yarn, and the development on a semi-works scale of the chemical engineering data for the erection and operation of a large-scale plant." [5] After many trial runs, yarn for an experimental

[5] E. Bolton, "Development of Nylon," *Industrial and Engineering Chemistry*, January 1942, p. 6.

batch of stockings was available in April 1937. In July 1938, a pilot plant was completed. In October 1938, DuPont announced its intention of building a commercial plant with an annual capacity of three million pounds, but before the plant was completed its capacity was increased to eight million pounds. The R and D leading to nylon was very expensive and time consuming. Eleven years elapsed from the beginning of research on superpolymers to the production of nylon on the first commercial unit. Although there is some disagreement over the size of the R and D expenditures, even the lowest estimate is about $2 million. According to a representative from Imperial Chemical Industries, about $800,000 was spent prior to the construction of the pilot plant, about $400,000 was spent on the pilot plant, and about $800,000 was spent on sales development. Other estimates of the costs are much higher.

Nylon provides an interesting illustration of the uncertainties involved in research and development. The original fiber was discovered quite by accident, and the project was almost dropped prior to completion. Also, this case illustrates the large investment that is sometimes required to carry an R and D project to completion. It should be recognized, however, that not all projects are as far-reaching as this. Most industrial R and D is directed at much more modest goals.[6]

## 4. The Process of Invention

"Invention" has been defined in many ways. According to one definition, an invention is a prescription for a new product or process that was not obvious to one skilled in the relevant art at the time the idea was generated. Other definitions add the requirement that the product or process must have prospective

[6] For further discussion of the research and development leading to nylon, see Jewkes, Sawers, and Stillerman, *op. cit.;* and W. Mueller, "The Origin of the Basic Inventions Underlying DuPont's Major Product and Process Innovations, 1920 to 1950," *The Rate and Direction of Inventive Activity*, National Bureau of Economic Research, 1962.

utility as well as novelty. This raises difficult questions as to how one is to find out whether a particular new product or process is prospectively useful, but it has the advantage of eliminating tinkering of an economically irrelevant sort. Thus, we include prospective utility as part of the definition.

Inventions can occur in either the research phase or the development phase of organized R and D activity. Generally, according to officials of the National Science Foundation, the central ideas come from research, and inventions in patentable form arise in the course of development. In addition, of course, many inventions occur through the efforts of independent inventors. How do inventions come about? According to one school of thought, they are due to the inspirations of genius, these inspirations not being susceptible to analysis. According to another school of thought, invention proceeds under the stress of necessity. If the great inventive geniuses had never lived, the same inventions would have been made by others without serious delay. When the time is ripe they are inevitable. Neither of these views is taken very seriously at the present time. The first is useless because the emergence of inventions is regarded as inexplicable. The second minimizes the significance of individual effort, ignores chance elements, and is too mechanistic.

Most economists view the situation differently. To them, invention is an activity characterized by great uncertainty but one which nonetheless shares most of the characteristics of other economic activities. In particular, they hypothesize that the amount of resources devoted to inventing in a particular field is dependent both on the social demand for inventions of this type and on the prospective costs of making the invention, the latter being related to the state of scientific knowledge. This, of course, is a variant of the conventional theories of expected profit or utility maximization. However, there is no intention of characterizing the inventor as an "economic man." It is recognized that, besides having economic motives, inventors invent for fun, fame, and the service of mankind, and perhaps to express the "instinct of workmanship" or the "instinct of contrivance." [7]

[7] See J. Schmookler, *Invention and Economic Growth*, Cambridge, Mass.: Harvard University Press, 1966; F. Taussig, *Inventors and Money Makers*,

It is also recognized that invention is inherently a very difficult process to analyze, map out, organize, and direct. Contrary to popular impression, this process frequently moves from the observation of a phenomenon to exploration of a use for it, not from a clearly defined goal to the discovery of technical means to achieve this goal. The process "does not usually move in a straight line, according to plan, but takes unexpected twists and turns." [8] Need and technique interact with one another, and it is not always apparent ahead of time from what disciplines or technologies answers will come. For example, a process for drawing brass rod arose from an adaptation of candle-making technology. Moreover, in trying to solve one problem, an answer to quite a different problem may result. Thus, Avicel, the non-nutritive food, was invented by an American Viscose Company chemist in the course of attempting to produce stronger rayon tire cord.

Turning to the psychological aspects of the process, some, like Abbot Usher,[9] postulate the existence of four steps leading to a successful invention. First, a problem of some kind is perceived. Second, the elements or data necessary for solving the problem are assembled through some particular set of events or train of thought, one of the elements being an individual with the required skill in manipulating the other elements. Third, an act of insight occurs, in which the solution of the problem is found. Fourth, there is a period of critical revision in which the solution becomes more fully understood and is worked into a broader context. To illustrate what Usher means by an act of insight, consider James Watt's most fundamental invention in the field of steam engine technology—the separate condenser. In 1763, a small model of the Newcomen steam engine was brought for repairs to Watt, a twenty-eight-year-old mathematical instrument maker at the University of Glasgow. Watt was perplexed by sev-

New York: The Macmillan Company, 1915; and T. Veblen, *The Instinct of Workmanship and the State of the Industrial Arts,* New York: The Macmillan Company, 1914.

8 See D. Schon, *Technology and Change,* New York: Delacorte Press, 1967, Chapter I.

9 A. Usher, *A History of Mechanical Inventions,* Cambridge, Mass.: Harvard University Press, 1954.

eral aspects of the model's operation. After some experimentation, he recognized a deficiency in the engine's concept. Then while strolling on the green of Glasgow "on a fine Sabbath afternoon" early in 1765, "the idea came into my mind that as steam was an elastic body it would rush into a vacuum, and if a connection were made between the cylinder and an exhausting vessel it would rush into it and might there be condensed without cooling the cylinder . . . I had not walked farther than the golf house, when the whole thing was arranged in my mind."[10]

Finally, a few words should be added regarding the characteristics of successful inventors. A study[11] based on a random sample of about 100 persons granted patents in 1953 indicates that about 50 percent were college graduates and that about 60 percent were technologists—engineers, chemists, metallurgists, and directors of research and development. Another study[12] investigates the age of inventors when they made "very important" inventions. In proportion to the number of inventors alive at various ages, very significant inventions were made at the highest average rate when inventors were not more than thirty to thirty-four years old. Moreover, the mean age of the inventors of 554 great inventions was about thirty-seven years. Thus the most significant inventions seem to be largely the product of relatively young men.

## 5. Expenditures on Research and Development: Growth and Flow of Funds

At this point we should provide a brief description of the basic statistics regarding R and D expenditures in the United States.

[10] Quoted in *ibid.*, p. 71.

[11] J. Schmookler, "Inventors Past and Present," *Review of Economics and Statistics*, August 1957. Also, see J. Rossman, *The Psychology of the Inventor*, Inventors Publishing Company, 1931.

[12] H. Lehman, *Age and Achievement*, Princeton, N.J.: Princeton University Press, 1953.

Recent decades have witnessed a tremendous growth in the amount spent on research and development. In 1945, industry performed about $1.3 billion worth of R and D; by 1968, this figure had increased to about $16.6 billion. In 1945, about $400 million of R and D was performed by government; by 1968, it performed about $3.5 billion of R and D. In 1945, the universities and other non-profit organizations performed about $200 million of R and D; by 1968 they performed about $4.2 billion. Table 3.1 shows the enormous increase in total R and D expenditures in the United States during the forties and fifties.[13] Paralleling this increase in R and D expenditures, there has been a great increase in the number of engineers and scientists engaged in research and development. In 1941, there were less than 90,000; in 1961, there were almost 400,000 (Table 3.1). Although the number of engineers and scientists engaged in R and D has increased at an impressive rate, it has not increased as rapidly as R and D expenditures. This is because the increases in demand for research personnel have resulted in higher salaries, and because less skilled labor and equipment seem to have been substituted where possible for the time of engineers and scientists.

It is also important to note that much of the R and D *performed* by one sector is *financed* by another sector. Table 3.2 shows that a large percentage of the R and D performed in the industrial sector is financed by the Federal government. In 1953, about 40 percent was financed in this way; in 1968, it had risen to about 50 percent. The situation is similar in the university sector. In 1953, about 60 percent of the R and D performed by

[13] The 1945 figures in the text come from D. Keezer, D. Greenwald, and R. Ulin, "The Outlook for Expenditures on Research and Development During the Next Decade," *American Economic Review,* May 1960. They are not entirely comparable with the figures in the tables below.

Of course, part of the increase in R and D expenditures is undoubtedly due to inflation and shifting definitions of R and D, but it is generally agreed that if these factors could be taken into account, there still would be a tremendous growth in R and D expenditures.

Note too that, although R and D performance in manufacturing increased considerably during the sixties, the ratio of R and D expenditures to sales of R and D performers increased only from 4.2 percent in 1960 to 4.4 percent in 1964. See "Basic Research, Applied Research, and Development in American Industry, 1964," *Review of Data on Science Resources,* National Science Foundation, January 1966. Also, the increase in the Federal R and D budget has slowed down. See Chapter VI.

*TABLE 3.1 Total R and D Expenditures and Number of Research Scientists and Engineers, United States, 1941–1968*

| *Year* | *Total R and D Expenditures (millions of dollars)* | *Number of Research Scientists and Engineers (thousands)* |
|---|---|---|
| 1941 | 900 | 87 |
| 1943 | 1,210 | 97 |
| 1945 | 1,520 | 119 |
| 1947 | 2,260 | 125 |
| 1949 | 2,610 | 144 |
| 1951 | 3,360 | 158 |
| 1953 | 5,160 | 223 [b] |
| 1955 | 6,200 | n.a. |
| 1957 | 9,810 | 327 [c] |
| 1959 | 12,430 | n.a. |
| 1961 | 14,380 | 387 |
| 1968 [a] | 24,330 | n.a. |

Source: The Growth of Scientific Research and Development, *U.S. Department of Defense, 1953, pp. 10 and 12;* National Science Foundation Review of Data on Research and Development, *No. 33, April 1962; and* National Science Foundation, Databook, *February 1969.*

[a] Preliminary.
[b] 1954 figure.
[c] 1958 figure.
n.a. Not available.

the universities was financed by the government; in 1968, it was about 55 percent. In addition, the Federal government financed a large and relatively stable (about 60 percent) portion of the R and D carried out by nonprofit organizations other than universities. Besides this massive outflow of funds from the Federal government to support R and D performed in other sectors, there were other flows of funds that are noteworthy, although much smaller. In recent years, industry financed about 2 percent of the R and D carried out by universities and about 8 percent of the R and D carried out by nonprofit organizations other than universities. Nonprofit organizations, such as the private foundations, financed about 4 percent of the R and D performed by colleges and universities.

*TABLE 3.2 Sources of R and D Funds and Performers of R and D, by Sector, United States, 1953 and 1968*

| Sources of R and D Funds (sector) | R and D Performance, by Sector: Federal Government | Industry | Colleges and Universities | Other Nonprofit Organizations | Total |
|---|---|---|---|---|---|
| *1953 Transfer of Funds (millions of dollars)* | | | | | |
| Federal government | 1,010 | 1,430 | 260 | 60 | 2,760 |
| Industry | | 2,200 | 20 | 20 | 2,240 |
| Colleges and universities | | | 120 | — | 120 |
| Other nonprofit organizations | | | 20 | 20 | 40 |
| Total | 1,010 | 3,630 | 420 | 100 | 5,160 |
| *1968 Transfer of Funds [a] (millions of dollars)* | | | | | |
| Federal government | 3,525 | 8,300 | 2,150 | 620 | 14,595 |
| Industry | | 8,300 | 55 | 70 | 8,425 |
| Colleges and universities | | | 944 | — | 944 |
| Other nonprofit organizations | | | 117 | 250 | 367 |
| Total | 3,525 | 16,600 | 3,266 | 940 | 24,331 |

Source: National Science Foundation Databook, *February 1969.*

[a] Preliminary figures.

## 6. Interindustry Differences in Expenditures

Which industries spend most on R and D? Which spend least? How much goes for basic research? Applied research? Development? How much goes for new products? Product improvements? New processes? These questions are obviously basic ones. Fortunately, the National Science Foundation's annual surveys of American industry help to answer many of them.

For many years, R and D performance as a percentage of sales has been highest in the aircraft, electrical equipment, instrument, and chemical industries (Table 3.3). Of course, this is due in considerable part to the fact that these industries carry out a great deal of R and D for the Federal government. In 1967, the Federal government financed about 80 percent of the R and D in the aircraft industry, 60 percent of the R and D in the electrical equipment industry, and 30 percent of the R and D in the

*TABLE 3.3 Performance of Industrial Research and Development, by Industry, 1927–1961*

| *Industry* | *1927* | *1937* | *1951* | *1957* [a] | *1961* [a] |
|---|---|---|---|---|---|
| | *(percent of sales)* | | | | |
| Aircraft and parts | n.a. | n.a. | 11.90 | 18.9 | 24.2 |
| Instruments | n.a. | n.a. | 3.00 | 7.3 | 7.3 |
| Electrical equipment | 0.54 | 1.50 | 3.60 | 11.0 | 10.4 |
| Chemicals | 0.42 | 1.10 | 1.50 | 3.5 | 4.6 |
| Rubber | 0.36 | 0.96 | 0.50 | n.a. | 2.2 |
| Machinery | 0.19 | 0.43 | 0.50 | 4.2 | 4.4 |
| Stone, clay, and glass | 0.13 | 0.43 | 0.40 | n.a. | 1.8 |
| Motor vehicles | 0.07 | 0.40 | 0.50 } | 2.9 } | 2.9 |
| Other transportation equipment | 0.07 | 0.07 | 0.30 } | | |
| Primary metals and products | 0.07 | 0.17 | n.a. | n.a. | n.a. |
| Fabricated metal | n.a. | n.a. | 0.30 | 1.5 | 1.3 |
| Primary metal | n.a. | n.a. | 0.20 | 0.5 | 0.8 |
| Petroleum | 0.09 | 0.45 | 0.70 | 0.8 | 1.0 |
| Paper | 0.06 | 0.17 | 0.30 | 0.7 | 0.7 |
| Food | 0.02 | 0.04 | 0.10 | 0.3 | 0.3 |
| Forest products | 0.01 | 0.04 | 0.03 | n.a. | 0.5 |
| Leather | 0.01 | 0.02 | 0.03 | n.a. | n.a. |
| Textiles and apparel | 0.01 | 0.02 | 0.07 | n.a. | 0.6 |

Source: *Y. Brozen, "Trends in Industrial Research and Development,"* Journal of Business, *July 1960, Table 3, and* Research and Development in Industry, 1961, *National Science Foundation, 1964.*

[a] The 1957 and 1961 figures are not entirely comparable with the earlier ones. See the *Source.*

n.a. Not available.

instruments industry. The situation in all industries in 1967 is shown in Table 3.4. When company-financed R and D rather than R and D performance is considered, the differences among industries are reduced but the industries remain in much the same rank order—instruments, electrical equipment, chemicals, and machinery being highest. Basic research constitutes the largest percentage of total R and D performance in the chemical (including drug) and petroleum industries, but in no case does it exceed 15 percent of the total. Development is a particularly large percentage of the total in the aircraft, electrical equipment, and machinery industries. For all industries combined, about 4 percent of the total is basic research, 18 percent applied research, and 78 percent development. Thus, the bulk of the funds goes for development, not research.

A survey of business plans for new plant and equipment provides further information regarding the character of the R and D being carried out by industry. In all manufacturing industries combined, about 47 percent of the firms reported in 1962 that their main purpose was to develop new products, 40 percent reported that it was to improve existing products, and 13 percent reported that it was to develop new processes. Development of new products seemed to be particularly important in the electrical equipment, chemical, and fabricated metal industries. Improvement of existing products seemed to be particularly important in the transportation equipment, machinery, auto, steel, and textile industries. Development of new processes was particularly important in the petroleum and rubber industries.[14] Finally, a considerable amount of the applied research and development performed in one industry is directed at products in another industry. In large part this is because firms classified in one industry are often in other industries as well. For example, many large petroleum refiners are chemical producers. In addition, it is because firms look somewhat beyond their own product lines for profitable R and D projects. An important moral is that one should not assume that all—or almost all—of the R and D performed in an industry is directed at its own products.

[14] These figures pertain to 1962, but it seems doubtful that there has been much of a change since then.

*TABLE 3.4 Research and Development Performance and Amount Financed by Federal Government, by Industry, 1967*

| Industry | *R and D Performance* | *Amount Financed by Federal Government* |
|---|---|---|
| | *(millions of dollars)* | |
| Food and kindred products | 168 | 1 |
| Paper and allied products | 94 | n.a. |
| Chemicals and allied products | 1,565 | 212 |
| Industrial chemicals | 1,006 | 182 |
| Drugs and medicines | 354 | n.a. |
| Other chemicals | 206 | n.a. |
| Petroleum refining and extraction | 469 | 38 |
| Rubber products | 195 | 30 |
| Stone, clay, and glass products | 152 | 7 |
| Primary metals | 245 | 8 |
| Primary ferrous products | 144 | 3 |
| Nonferrous and other metal products | 102 | 5 |
| Fabricated metal products | 165 | 12 |
| Machinery | 1,438 | 393 |
| Electrical equipment and communication | 3,806 | 2,240 |
| Communication equipment and electronic components | 2,241 | 1,341 |
| Other electronic equipment | 1,566 | n.a. |
| Motor vehicles and other transportation equipment | 1,377 | 389 |
| Aircraft and missiles | 5,568 | 4,510 |
| Professional and scientific instruments | 464 | 147 |
| Scientific and mechanical measuring instruments | 78 | 14 |
| Optical, surgical, photographic, and other instruments | 385 | 133 |
| Textiles | 52 | n.a. |
| Lumber | 14 | — |

Source: National Science Foundation Reviews of Data on Science Resources, *No. 17, February 1969.*

n.a. Not available.

We must look more closely at the large interindustry differences in the ratio of company-financed R and D expenditures to sales and investigate the reasons why these differences exist.[15] At least three factors seem to be important. First, industries differ considerably in the value their customers place on increased performance. Being second best in product performance in some fields is not a great handicap because consumers do not care very much about the difference in performance; in other industries, a second-best product has relatively little value. For example, improvements in drugs seem to have great importance for consumers in the sense that they are willing to pay more for a more effective medicine. In industries where this is the case, it is not surprising that the ratio of R and D expenditures to sales tends to be high. Second, industries seem to differ considerably in the ease with which research and development can bring about significant inventions. Although an industry's "science base" is a very slippery concept, it is difficult to deny that some industries, like electronics and chemicals, lie closer to well-developed basic sciences than others and that for this reason the effectiveness of a given amount of research and development may be greater than in other industries. Third, industries differ in market structure. Industries composed of many very small firms are unlikely to spend as much on R and D as somewhat less fragmented industries do. More will be said on this score in section 12.

Of course, the fact that some industries spend much more than others on R and D does not mean that the latter are spending too little. As we noted in Chapter II, an industry's rate of technological change depends on the nature and extent of the research and development carried out by other industries as well as its own. Many industries, particularly those producing consumer goods, rely on their suppliers and equipment producers to carry out research and development. In effect, they buy R and D, incorporated in new products, from their suppliers. Other industries, like agriculture, medicine, and aviation, rely in an important way on research and development carried out by govern-

[15] Chapter V discusses the factors influencing the level and allocation of government R and D expenditures.

ment laboratories. There are advantages in certain industries specializing more than others in research and development. Whether this specialization has gone too far is hard to say; the issues are discussed at some length in Chapter VI.

## 7. Uncertainty in Research and Development

Chance plays a crucial role in research and development, and a long string of failures often occurs before any sort of success is achieved. For example, a survey of 120 large companies doing a substantial amount of R and D indicates that, in half of these firms, at least 60 percent of the R and D projects never resulted in a commercially used product or process. (The smallest failure rate for any of these firms was 50 percent.) Moreover, even when a project resulted in a product or process that was used commercially, the profitability of its use was likely to be quite unpredictable.[16]

A study[17] carried out by the RAND Corporation goes further in describing the extent of the difficulties in predicting the results of military development projects. First, the RAND study showed that there were substantial errors in the estimates (made prior to development) of the costs of producing various types of military hardware. When adjusted for unanticipated changes in factor prices and production-lot sizes, the average ratio of the actual to estimated cost was 1.7 (fighters), 3.0 (bombers), 1.2 (cargoes and tankers), and 4.9 (missiles). Thus, the estimates made prior to the development of these types of equipment were, on the average, in error by several hundred percent—and almost always they understated the actual costs.

Second, the extent to which costs were understated was directly related to the extent of the technological advance. In cases where

[16] *Chemical and Engineering News,* July 10, 1957. For estimates of the proportion of products emerging from research and development that fail at various stages see Chapter IV.

[17] A. Marshall and W. Meckling, "Predictability of the Costs, Time, and Success of Development," *The Rate and Direction of Inventive Activity,* Princeton, N.J.: Princeton University Press, 1962.

a "large" technological advance was required, the average ratio was 4.2; in cases where a "small" technological advance was required, the average ratio was 1.3. Moreover, when corrected for bias, there was much more variation in the ratio in cases where the required technological advance was large than in those where it was small. Of course, this is what one would expect.

Third, there were very substantial errors in the estimated length of time it would take to complete a project. For ten weapons systems, the average error was two years, the maximum being five years. The average ratio of the actual to the expected length of time was 1.5, indicating once again that estimates tend to be overly optimistic. The results suggest too that the estimates are more accurate when "small" technological advances are required than when "large" technological advances must be made. Fourth, given the extent of the technological advance that had to be made, the estimates of costs and development time became more accurate as the project ran its course. For example, at the early stages of projects requiring advances of "medium" difficulty, the average ratio of the actual to expected cost was 2.15 and the standard deviation[18] was .57. At the middle stages of such projects, the average ratio was 1.32 and the standard deviation was .39. At the late stages of such projects, the average ratio was 1.06 and the standard deviation was .18.

These findings pertain entirely to military R and D. To some extent, they reflect the fact that defense contractors have had an incentive in the past to make optimistic estimates, the penalties for being over-optimistic often having been small and the possible rewards having been large. They also reflect the ambitious nature of development in the defense sphere. Nonetheless they are not an utterly undependable guide to the civilian economy. The errors in estimation in the civilian economy, although smaller than those presented above, are also quite large, particularly when large technological advances are attempted.[19]

[18] The standard deviation, a measure of dispersion, is the square root of the mean squared deviation of observations from their mean.

[19] E. Mansfield, *Industrial Research and Technological Innovation, op. cit.,* Chapter III.

## 8. Learning and Parallel Research and Development Efforts

A research or development project can be regarded as a process of uncertainty reduction, or learning. Suppose, for example, that a firm which is trying to fabricate a part can use one of two alloys and that it is impossible to use standard sources to determine their characteristics. Suppose that strength is of paramount importance and that the firm's estimates of the strengths of the

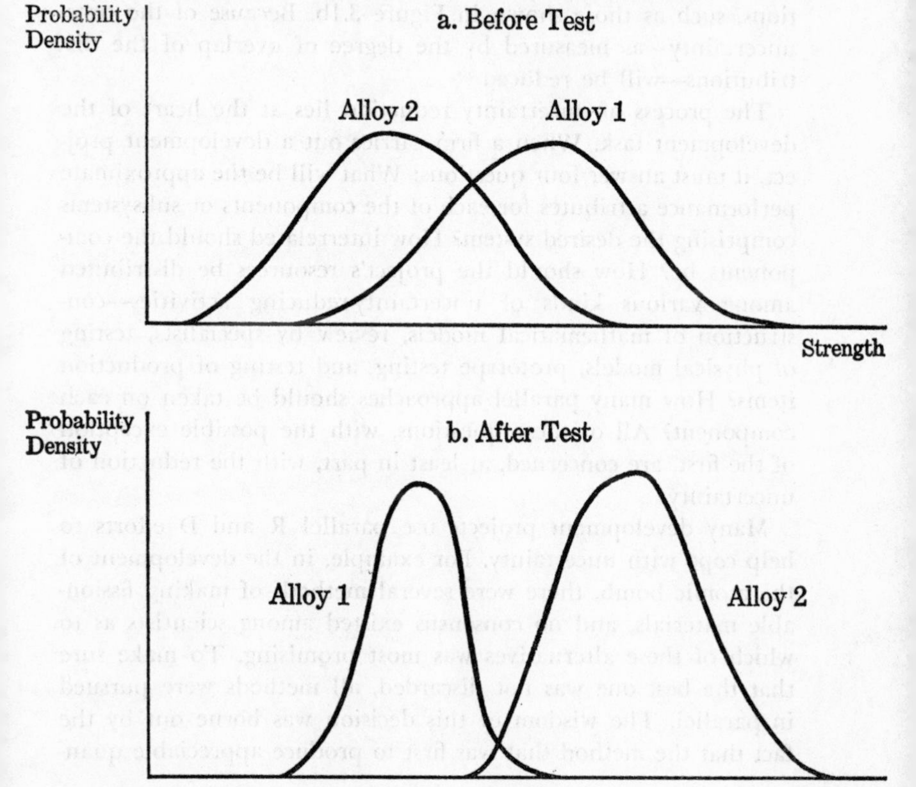


*FIGURE 3.1 Subjective Probability Distributions of Strength of Alloys 1 and 2, before and after Test*

alloys—Alloy 1 and Alloy 2—are represented by the subjective probability distributions in Figure 3.1a. If the firm were forced to make a choice immediately, it would probably choose Alloy 1, since it believes there is better than a fifty-fifty chance that Alloy I will turn out to be stronger than Alloy 2. However, there is a good chance that this decision would be wrong, the consequence being that the part would be weaker than if Alloy 2 had been used. Thus, the firm may decide to perform a test prior to making the selection. On the basis of the test results, the firm will formulate new estimates represented by new probability distributions, such as those shown in Figure 3.1b. Because of the tests, uncertainty—as measured by the degree of overlap of the distributions—will be reduced.[20]

The process of uncertainty reduction lies at the heart of the development task. When a firm carries out a development project, it must answer four questions: What will be the approximate performance attributes for each of the components or subsystems comprising the desired system? How interrelated should the components be? How should the project's resources be distributed among various kinds of uncertainty-reducing activities—construction of mathematical models, review by specialists, testing of physical models, prototype testing, and testing of production items? How many parallel approaches should be taken on each component? All of these questions, with the possible exception of the first, are concerned, at least in part, with the reduction of uncertainty.

Many development projects use parallel R and D efforts to help cope with uncertainty. For example, in the development of the atomic bomb, there were several methods of making fissionable materials, and no consensus existed among scientists as to which of these alternatives was most promising. To make sure that the best one was not discarded, all methods were pursued in parallel. The wisdom of this decision was borne out by the fact that the method that was first to produce appreciable quan-

[20] This example is similar to one presented in T. Marschak, T. Glennan, and R. Summers, *Strategy for R and D,* New York, N.Y., Springer-Verlag, 1967. See any elementary textbook in statistics for a discussion of probability distributions.

tities of fissionable material was one that had been considered relatively unpromising early in the development program.

Under what conditions is it optimal to run parallel R and D efforts? What factors determine the optimal number of parallel efforts? Suppose that an R and D manager can select $n$ approaches, spend $M$ dollars on each one over a period of $\theta$ months, pick the one that looks most promising at the end of the period, and follow it to completion, dropping the others. Suppose that the only relevant criterion is the extent of the development costs, the usefulness of the result and the development time being assumed to be the same regardless of what strategy is pursued. For further simplification, suppose that all approaches look equally promising and that the results of the approaches are independent.

Under these conditions, the optimal value of $n$—the number of parallel R and D efforts—is inversely related to $M$ and directly related to the amount learned in the next $\theta$ months. This result can be proved easily, but it requires some mathematics. Intuitively, it is eminently plausible. As the cost of running each effort increases, one would certainly expect the optimal number of parallel efforts to decrease. Moreover, as the prospective amount of learning increases, one would also expect the optimal number of parallel efforts to increase. This result is important because it puts in proper perspective some of the criticisms of "duplication" and "waste" in research and development. Contrary to popular belief, parallel efforts may produce results more quickly and more cheaply than attempting in advance to choose the optimal approach and concentrating all one's efforts on pursuing it. The fact that most of the parallel paths are ultimately rejected does not mean that they are a waste. On the contrary, given considerable uncertainty, this may be the cheapest way to proceed.[21]

[21] See R. Nelson, "Uncertainty, Learning, and the Economics of Parallel Research and Development Efforts," *Review of Economics and Statistics*, 1961. Unfortunately, difficulties arise when one attempts to extend the solution to cases involving more than one decision point. See Marschak, Glennan, and Summers, *op. cit.* Also see F. Scherer, "Time-Cost Tradeoffs in Empirical Research Projects," *Naval Logistics Research Quarterly*, 1966.

## 9. Determinants of Development Costs

What determines how much a particular development project will cost? First, there is the size and complexity of the product being developed. It takes more resources to redesign a big product with a large number of components, because there are more drawings to be made, more analyses to be done, and more tests to perform. Moreover, holding the number of components constant, the cost of developing a product tends to be directly related to the interdependence among these components. If there is considerable interdependence, a change in one part requires the redesign of other parts. Thus, the cost of making a major alteration tends to be high.

Second, there is the extent of the advance in performance that is sought. Larger advances generally mean higher costs because more components will need to be specially designed, more mistakes will be made, and more jobs will have to be redone. When larger advances are sought, there is generally more uncertainty surrounding the project; in extreme cases, it may even be difficult to tell which technological problems will be most important. Costs must be incurred to reduce the uncertainties; the project tends to be more expensive, if possible at all, than if it attempts a more modest advance in the state of the art.

Third, there is development time. Some economists have postulated that there exists a function—like that in Figure 3.2—which relates expected development time to the expected total quantity of resources employed in a development effort. Their principal point is that between *N* and *R* the time-cost function has a negative slope, indicating that time can be decreased only by increasing total cost. As the development schedule is shortened, more tasks must be carried out concurrently rather than sequentially, and since each task provides knowledge that is useful in carrying out the others, there are more false starts and wasted designs. Also, diminishing returns set in as more and more technical workers are assigned simultaneously to the development effort. It is argued that these factors offset the possibility that shorter development time eliminates unnecessary work and some overhead-type costs.

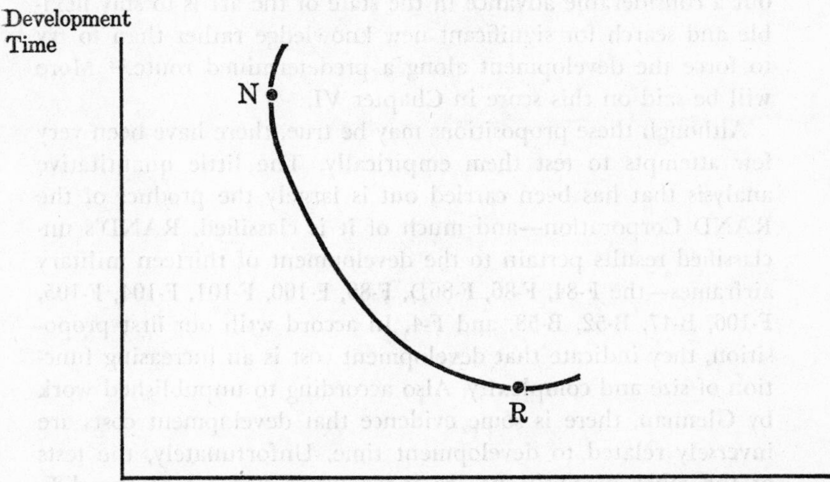


*FIGURE 3.2 Development Possibility Curve*

Source: *F. M. Scherer, "Government Research and Development Programs,"* Measuring Benefits of Government Expenditures, *edited by R. Dorfman, Washington, D.C.: The Brookings Institution, 1965.*

Fourth, there is the stock of basic knowledge and of components and materials. The task of developing a particular product can be made much easier and cheaper by advances in basic knowledge which allow better prediction of relevant phenomena. Similar effects are produced by advances in the quality of test equipment (and such changes as the greater use of computers in design). Moreover, the development task is obviously made easier by improvements in components and materials, improvements in materials often being particularly important.

Fifth, there is the development strategy that is used. Essentially, the development strategy is the set of decision rules employed to allocate resources during the course of the project. A great deal of attention has been devoted to development strategy, particularly with regard to the development of military aircraft. It is often argued that the proper strategy when trying to carry out a considerable advance in the state of the art is to stay flexible and search for significant new knowledge rather than to try to force the development along a predetermined route.[22] More will be said on this score in Chapter VI.

Although these propositions may be true, there have been very few attempts to test them empirically. The little quantitative analysis that has been carried out is largely the product of the RAND Corporation—and much of it is classified. RAND's unclassified results pertain to the development of thirteen military airframes—the F-84, F-86, F-86D, F-89, F-100, F-101, F-104, F-105, F-106, B-47, B-52, B-58, and F-4. In accord wtih our first proposition, they indicate that development cost is an increasing function of size and complexity. Also according to unpublished work by Glennan, there is some evidence that development costs are inversely related to development time. Unfortunately, the tests of the other propositions are inconclusive because of the difficulty of measuring the relevant factors and because of the smallness of the sample.[23]

## 10. The Importance of the Independent Inventor

In the last three sections of this chapter, we consider the role of the independent inventor in promoting technological change, as well as certain aspects of the role of the large firm. Although the growth of organized research and development has led to many attempts to administer the last rites to the independent inven-

[22] See B. Klein, "Policy Issues Involved in the Conduct of Military Development Programs," *The Economics of Research and Development,* edited by R. Tybout, Columbus, Ohio: Ohio State University Press, 1965; and "The Decision Making Problem in Development," *The Rate and Direction of Inventive Activity,* National Bureau of Economic Research, 1962.

[23] T. Glennan, "Methodological Problems in Evaluating the Effectiveness of Military Aircraft Development," RAND Corporation P-3357, May 1966.

tor, he is by no means dead. Over the last sixty years he has contributed a great deal to the stream of important inventions. For example, in their study of sixty-one significant twentieth-century inventions, Jewkes, Sawers, and Stillerman [24] estimated that over half were produced by individuals not doing company-directed research. Nonetheless, this century has seen a notable shift in the source of inventions away from the independent inventor and toward the corporation. In 1900, about 80 percent of all patents (issued to individuals and United States firms) were issued to individuals; in 1957, about 40 percent were issued to individuals. The reasons for this shift are not difficult to find. Technology in most industries has become more complex, a division of labor among specialists in various scientific fields has become more necessary, and the instruments required to research and develop new processes and products have become more expensive.

Despite these factors, the independent inventor, the inventor who works outside industrial R and D organizations, continues to be an important source of new technology. The independent inventor tends to move into areas where the costs of inventing are low, and where time and ingenuity can be substituted for expensive equipment. For example, Jacob Rabinow invested only about $20,000 in his invention of the clock regulator. The independent inventor sometimes is willing to undertake projects that company R and D is not imaginative enough to pursue. For example, it was an independent inventor, Frank Whittle, who, without the support of the British Air Ministry or aircraft industry, contributed so much to the development of the early jet engine. Incidentally, it is by no means true that all independent inventors are uneducated tinkerers. On the contrary, some are highly trained scientists and engineers, sometimes university faculty members. Examples are Leo Baekeland, who invented Bakelite, Eugene Houdry, who invented catalytic cracking, Edwin Armstrong, who invented FM radio broadcasting, and Edwin Land, who invented the Polaroid camera.

Independent inventors often turn their inventions over to larger companies for development, because they lack the financial

[24] J. Jewkes, D. Sawers, and R. Stillerman, *op. cit.*

resources and necessary facilities. This was true in the case of Jacques Brandenberger, a French chemist, who is generally credited with the invention of cellophane. The Comptoir de Textiles Artificiels, the biggest French rayon producer, became interested in his results and agreed to finance the development work. A new company, La Cellophane, was formed. Brandenberger's patents were transferred to it, and Brandenberger was employed to direct the development. La Cellophane was the first commercial manufacturer of plain cellophane. Turning to personal characteristics, independent inventors have the reputation of being a stubborn and nonconforming breed. Many observers claim that the successful independent inventor often is marked by a persistence bordering on obsession and that he tends to be more of a maverick than his corporate counterpart.

Xerography provides a further illustration of the work of an independent inventor. After graduating from Caltech in 1930, Chester Carlson worked at the Bell Telephone Laboratories, where he transferred to the patent department and studied patent law. In 1934, he took a job as patent attorney at Mallory and Company. Noticing how difficult and costly it was to copy documents, he set out to examine the problem. By 1937, he had developed the basic concept of xerography and filed a patent application although he had yet to verify his ideas experimentally. After several years of improving his process, he went to the Battelle Memorial Institute, the world's largest nonprofit research organization, and made an agreement whereby Battelle acquired the patent rights and Carlson acquired a substantial share in the proceeds of any future development. Battelle improved the process to the point where industry became interested. In 1946, the Haloid Corporation (later renamed the Xerox Corporation) accepted an exclusive sublicense, developed the process further, and in 1950 announced its first commercial application.[25]

[25] See the testimony of D. DeSimone before the Senate Subcommittee on Antitrust and Monopoly, May 18, 1965.

## 11. The Sources of Radical Inventions

Some observers claim that the bulk of the commercial R and D carried out by large corporations is aimed at fairly modest advances in the state of the art.[26] They say that the really major inventions seldom stem from the laboratories of the large firms, which are primarily contributors of minor "improvement" inventions. If true, these hypotheses have important implications regarding the role of the large firm in promoting technological change.

The evidence cited in support of these hypotheses are of two types. First, the McGraw-Hill surveys indicate that about 90 percent of the firms expected their R and D outlays to pay off in five years or less. Since it usually takes considerably longer than this before a radically new process or product even hits the market, the emphasis on short pay-off periods is taken to indicate that most R and D in these firms is geared toward improvements or minor changes in existing products. Moreover, the stated objectives of the research programs of various firms indicate an emphasis on immediate and short-term benefits. Second, studies of the origins of various major twentieth-century inventions seem to indicate that only a fairly small proportion originated in the laboratories of large corporations. Jewkes, Sawers, and Stillerman[27] found that such laboratories accounted for only twelve of the sixty-one major inventions they studied. Thirty-three of the inventions—including air conditioning, automatic transmission, Bakelite, cellophane, and the jet engine—were the product of independent inventors. Hamberg[28] found that the labora-

[26] Of course, no one denies that the risks are greater than in most other aspects of business. Thus, there is no disagreement with section 7. The comparison is with very far-reaching R and D projects. See D. Hamberg, "Invention in the Industrial Laboratory," *Journal of Political Economy*, April 1963, J. Jewkes, D. Sawers, and R. Stillerman, *op. cit.*, and R. Nelson M. Peck, and E. Kalachek, *Technology, Economic Growth and Public Policy*, Washington, D.C.: The Brookings Institution, 1967; and E. Mansfield, *op. cit.*, Chapter IV.

[27] J. Jewkes, D. Sawers, and R. Stillerman, *op. cit.*

[28] D. Hamberg, *op. cit.*

tories of large firms accounted for only seven of the twenty-seven major post-World War II inventions he studied. The rest—including the ENIAC computer, the oxygen converter, stereophonic sound, and neomycin—came from independent inventors, small firms, universities, and an agricultural experiment station.

There is probably a good deal of truth in these hypotheses, although they certainly do not hold for all firms. From the point of view of the individual large firm, it is often wise to leave the pioneering to others and to stick to less far-reaching research and development. Whether this is wise from the point of view of society is a different and more difficult question. Some economists assert that, from the point of view of society, large firms tend to spend too little on more fundamental, risky, and radical types of R and D. Even if this is the case (and it may well be), it does not deny the importance of the laboratories of large corporations. On the contrary, as we pointed out in Chapter II, technological change in many industries may be due in considerable measure to the cumulative effect of many "improvement" inventions. Although they may be a less important source of really major inventions than is socially desirable, their contribution is nonetheless of great significance. Whether or not very large firms are required to make this contribution is another question, and one which will be discussed below and in subsequent chapters.

## 12. Size of Firm, Market Structure, and Research and Development

In Chapter I, we pointed out that some economists believe that very large firms are needed to produce the technical achievements on which economic progress depends. For example, John Kenneth Galbraith has asserted that only such firms can carry out the activities required in the modern world to develop new products and processes; and on this basis, he has argued that an industry composed of a few large oligopolists will be technically more progressive than an industry composed of a large number of smaller firms. Various reasons have been given for the alleged importance of great size: a large firm can finance R and D more easily, it can afford bigger projects, the results of R and D are more likely to be useful because of its greater diversity, it can wait longer for the pay-off, and it can capture a larger portion of the social gains from its research because it has a larger share of the market.[29] Not all economists agree with Galbraith; many do not believe that very large firms are required to produce rapid technological change.[30]

In studying this issue, a relevant question is: do large firms spend more, relative to their size, than small firms on R and D? Although the National Science Foundation's data cast only a limited amount of light on this issue, it is worthwhile to see what they suggest. Table 3.5 shows the relationship in various industries between the amount spent on R and D as a percent of sales and a firm's size, for those firms that engaged in R and D. Looking at R and D performance, there seems to be a tendency in most industries for the largest firms to carry out relatively more R and D than the other firms. Only in lumber, drugs, and nonferrous metals is this not the case. Looking at company-financed R and D, there is a somewhat weaker tendency for the largest firms to support a larger amount of R and D (as a percent of sales) than smaller firms, but in most industries such a tendency exists. If data were available for all firms (not just those that perform R and D), this tendency would be strengthened, because a larger proportion of large firms than small ones perform R and D.

Thus, the results seem to favor the hypothesis that large firms support more R and D than do small ones. However, from the point of view of antitrust policy, the more relevant question is: Do the largest few firms in an industry spend more on R and D

[29] J. K. Galbraith, *American Capitalism,* Boston: Houghton Mifflin Company, 1952; J. Schumpeter, *Capitalism, Socialism, and Democracy,* New York: Harper & Row, Publishers, 1947; and H. Villard, "Competition, Oligopoly, and Research," *Journal of Political Economy,* December 1958.

[30] See J. Jewkes, D. Sawers, and R. Stillerman, *op. cit.;* G. Nutter, "Monopoly, Bigness, and Progress," *Journal of Political Economy,* June 1956; and E. Mansfield, *Monopoly Power and Economic Performance,* New York: W. W. Norton & Company, Inc., 1968. The latter brings together a variety of views on this subject.

*TABLE 3.5 Research and Development Performance and Company-Financed R and D Costs of Firms Performing R and D, by Industry and Size of Firm's Work Force, 1961*

| | *R and D Performance* | | | *Company-Financed R and D* | | |
|---|---|---|---|---|---|---|
| | *Employment* | | | *Employment* | | |
| *Industry* | *Under 1,000* | *1,000–4,999* | *5,000 or More* | *Under 1,000* | *1,000–4,999* | *5,000 or More* |
| | *(percent of sales)* | | | | | |
| Food | n.a. | 0.3 | 0.4 | n.a. | 0.3 | 0.4 |
| Textiles | n.a. | 0.5 | 0.5 | n.a. | n.a. | 0.4 |
| Lumber | 1.8 | 0.6 | 0.3 | n.a. | n.a. | 0.3 |
| Paper | n.a. | 0.7 | 0.7 | n.a. | 0.8 | 0.7 |
| Chemicals | 2.2 | 3.7 | 5.2 | n.a. | 3.4 | 3.9 |
| Industrial | 3.1 | 4.4 | 5.8 | n.a. | 3.2 | 4.7 |
| Drugs | 3.8 | 5.8 | 4.4 | n.a. | 6.4 | 4.6 |
| Others | 1.5 | 2.1 | 3.8 | n.a. | 2.1 | 1.6 |
| Petroleum | n.a. | 0.6 | 1.0 | n.a. | 0.6 | 1.0 |
| Rubber | 0.8 | 0.9 | 2.6 | n.a. | 0.9 | 1.7 |
| Stone, clay, and glass | n.a. | 1.1 | 2.1 | n.a. | 1.1 | 2.1 |
| Primary metals | n.a. | 0.9 | 0.8 | n.a. | 0.8 | 0.8 |
| Ferrous | n.a. | 0.4 | 0.7 | n.a. | 0.4 | 0.8 |
| Nonferous | n.a. | 1.3 | 1.0 | n.a. | 1.1 | 0.8 |
| Fabricated metal products | 1.0 | 1.0 | 1.7 | n.a. | 1.0 | 1.2 |
| Machinery | 3.0 | 2.0 | 6.2 | n.a. | 1.9 | 4.1 |
| Electrical equipment | 5.6 | 5.9 | 12.1 | n.a. | 2.8 | 4.2 |
| Motor vehicles | n.a. | 1.3 | 3.5 | n.a. | 0.9 | 2.7 |
| Aircraft | 6.9 | 12.6 | 25.7 | n.a. | 1.7 | 2.5 |
| Instruments | 4.8 | 4.3 | 9.6 | 1.7 | 3.3 | 5.3 |

Source: Research and Development in Industry, 1961, *National Science Foundation, 1964.*

n.a. Not available.

(as a percent of sales) than firms which are relatively large but still only a fraction of the size of the leaders? Table 3.5 cannot answer this question because, in most industries, the "smaller" firms as well as the giants have more than 5,000 employees, and, consequently, they are lumped together. Some light is thrown on this question by a study[31] that investigates the relationship between size of firm and the level of R and D expenditures among the major firms in the chemical, petroleum, drug, steel, and glass industries. Except for the chemical industry, the results provide no evidence that the largest firms in these industries spent more on R and D, relative to sales, than did somewhat smaller firms. In the petroleum, drug, and glass industries, the largest firms spent significantly less; in the steel industry, they spent less but the difference was not statistically significant. A study by Scherer provides similar results for a wider range of industries, the ratio of R and D expenditures to sales usually being no higher among the largest firms than among their somewhat smaller competitors.[32]

Needless to say, these findings pertain to only one of many questions relating to the effects of firm size and market structure on the rate of technological change. Little can be concluded from these results alone. In Chapter VI we bring together a variety of findings that bear on this subject.

[31] E. Mansfield, *op cit.*, Chapter II.

[32] See F. Scherer, Testimony Before the Senate Subcommittee on Antitrust and Monopoly, May 25, 1965, and "Size of Firm, Oligopoly, and Research: A Comment," *Canadian Journal of Economics and Political Science,* May 1965.

A difference is not statistically significant if the chances are better than one in twenty that it could be due to chance.

CHAPTER FOUR

# Innovation and the Diffusion of New Techniques

## 1. Innovation: Definition and Importance

An invention, when applied for the first time, is called an innovation. Traditionally economists have stressed the distinction between an invention and an innovation on the ground that an invention has little or no economic significance until it is applied. This distinction becomes somewhat blurred in cases like DuPont's nylon, where the inventor and the innovator are the same firm. Under these circumstances, the final stages of development may entail at least a partial commitment to a market test. However, in many cases the inventor is not in a position to —and does not want to—apply his invention, because his business is invention, not production, or because he is a supplier, not a user, of the equipment embodying the invention, or for some other reason. In these cases, the distinction remains relatively clear-cut.

Regardless of whether the break between invention and innovation is clean, innovation is a key stage in the process leading to the full evaluation and utilization of an invention. The innovator—the firm that is first to apply the invention—must be willing to take the risks involved in introducing a new and untried process, good, or service. In many cases, these risks are high. Although R and D can provide a great deal of information regarding the technical characteristics and cost of production of the invention—and market research can provide considerable information regarding the demand for it—there are many areas of uncertainty which can be resolved only by actual production and marketing of the invention. By obtaining needed information regarding the actual performance of the invention, the innovator plays a vital social role. In this chapter, we take up the nature of the innovative process and the characteristics of innovators, as well as the factors determining how rapidly, once an innovation occurs, its use spreads.

## 2. The Lag from Invention to Innovation

How long is the lag between invention and innovation? This lag must vary substantially, since some inventions require changes in taste, technology, and factor prices before they can profitably be utilized, whereas others do not. Moreover, some inventions constitute major departures from existing practice, whereas others are more routine "improvement" inventions.[1] Restricting our attention to relatively important inventions, the only data are extremely rough,[2] since concepts such as "invention" and "innovation" are not easy to pinpoint and date, and the available

[1] The distinction between "major" and "improvement" inventions is made by many writers. Tables 4.1 and 4.2 pertain to "major" inventions. The results regarding patent utilization (cited in Chapter VI) may be of use in describing the lag for "improvement" inventions.

[2] J. Enos, "Invention and Innovation in the Petroleum Refining Industry," *The Rate and Direction of Inventive Activity,* Princeton, N.J.: Princeton University Press, 1962; F. Lynn, "An Investigation of the Rate of Development and Diffusion of Technology in Our Modern Industrial Society," *Report of the National Commission on Technology, Automation, and Economic Progress,* Washington, D.C.: 1966.

*TABLE 4.1 Estimated Time Interval Between Invention and Innovation, Forty-Six Inventions, Selected Industries* [a]

| *Invention* | *Interval (years)* | *Invention* | *Interval (years)* |
|---|---|---|---|
| Distillation of hydrocarbons with heat and pressure (Burton) | 24 | DDT | 3 |
| Distillation of gas oil with heat and pressure (Burton) | 3 | Electric precipitation | 25 |
| Continuous cracking (Holmes-Manley) | 11 | Freon refrigerants | 1 |
| Continuous cracking (Dubbs) | 13 | Gyrocompass | 56 |
| "Clean circulation" (Dubbs) | 3 | Hardening of fats | 8 |
| Tube and tank process | 13 | Jet engine | 14 |
| Cross process | 5 | Turbojet engine | 10 |
| Houdry catalytic cracking | 9 | Long-playing record | 3 |
| Fluid catalytic cracking | 13 | Magnetic recording | 5 |
| Gas lift for catalyst pellets | 13 | Plexiglass, lucite | 3 |
| Catalytic cracking (moving bed) | 8 | Cotton picker | 53 |
| Safety razor | 9 | Nylon [b] | 11 |
| Fluorescent lamp | 79 | Crease-resistant fabrics | 14 |
| Television | 22 | Power steering [c] | 6 |
| Wireless telegraph | 8 | Radar | 13 |
| Wireless telephone | 8 | Self-winding watch | 6 |
| Triode vacuum tube | 7 | Shell molding | 3 |
| Radio (oscillator) | 8 | Streptomycin | 5 |
| Spinning jenny | 5 | Terylene, dacron | 12 |
| Spinning machine (water frame) | 6 | Titanium reduction | 7 |
| Spinning mule | 4 | Xerography | 13 |
| Steam engine (Watt) | 11 | Zipper | 27 |
| Ball point pen | 6 | Steam engine (Newcomen) | 6 |

Source: *J. Enos,* op. cit., *p. 307–308.*

[a] The first eleven inventions in the left-hand column were those that occurred in petroleum refining.

[b] Actually, this is the length of time between the beginning of fundamental research by DuPont on superpolymers and the production of nylon on the first commercial unit.

[c] This figure pertains to Vickers' booster units, not Davis's system.

samples are not random. Nonetheless, these data provide some feel for the distribution of the lag. John Enos estimated the time interval between invention and innovation for eleven important petroleum refining processes and thirty-five important products and processes in a variety of other industries. Table 4.1 shows that the lag averaged eleven years in the petroleum industry and about fourteen years in the others. Its standard deviation is about five years in the petroleum industry and sixteen years in the others. He concludes that: "Mechanical innovations appear

*TABLE 4.2 Average Rate of Development of Selected Technological Innovations* [a]

| *Factors Influencing the Rate of Technological Development* | *Average Time Interval (years)* | | |
|---|---|---|---|
| | *Incubation Period* [b] | *Commercial Development* [c] | *Total* |
| ***Time Period*** | | | |
| Early twentieth century (1885–1919) | 30 | 7 | 37 |
| Post-World War I (1920–1944) | 16 | 8 | 24 |
| Post-World War II (1945–1964) | 9 | 5 | 14 |
| ***Type of Market Application*** | | | |
| Consumer | 13 | 7 | 20 |
| Industrial | 28 | 6 | 34 |
| ***Source of Development*** | | | |
| Private industry | 24 | 7 | 31 |
| Federal government | 12 | 7 | 19 |

Source: *Frank Lynn, "An Investigation of the Rate of Development and Diffusion of Technology in Our Modern Industrial Society,"* Report of the National Commission on Technology, Automation, and Economic Progress, *Washington, D.C., 1966.*

[a] Based on study of twenty major innovations whose commercial development started in the period 1885–1950.

[b] Incubation Period—begins with basic discovery and establishment of technological feasibility, and ends when commercial development begins.

[c] Commercial Development—begins with recognition of commercial potential and the commitment of development funds to reach a reasonably well-defined commercial objective, and ends when the innovation is introduced as a commercial product or process.

to require the shortest time interval, with chemical and pharmaceutical innovations next. Electronic innovations took the most time. The interval appears shorter when the inventor himself attempts to innovate than when he is content merely to reveal the general concept."[3]

In a more recent study, Frank Lynn estimated the average number of years elapsing from the basic discovery and establishment of an invention's technical feasibility to the beginning of its commercial development, as well as the average number of years elapsing from the beginning of its commercial development to its introduction as a commercial product or process. The results, based on brief histories of twenty major innovations during 1885–1950, seem to indicate that the lag has been decreasing over time, that it is much shorter for consumer products than industrial products, and that it is much shorter for innovations developed with government funds than for those developed with private funds (Table 4.2).[4]

## 3. The Decision to Innovate

What factors should a firm consider in deciding whether or not to innovate? For present purposes, it is sufficient to provide a broad and highly simplified sketch, a detailed analysis being available elsewhere.[5] To begin with, the firm should estimate, of course, the expected rate of return from introducing the new product or process. In the case of a new product, the result obviously will depend on the capital investment that is required to introduce the innovation, the forecasted sales, the estimated costs of production, and the effects of the innovation on the costs

[3] J. Enos, *op. cit.*, p. 309.

[4] There are many pitfalls in this area. For example, even if there is no tendency for the lag to decrease over time, there can appear to be such a tendency because many recent inventions that have not yet been applied—and which experience long lags between invention and innovation—are necessarily omitted.

[5] For example, in E. Pessemier, *New-Product Decisions*, New York: McGraw-Hill, Inc., 1966.

and sales of the firm's existing product line. These factors depend in part on a firm's pricing policy, as well as on the characteristics of the new product. In addition, the firm should estimate, as best it can, the risks involved in innovating. These risks tend to be substantial, as witnessed by Booz, Allen, and Hamilton's estimate that, out of every ten products which emerge from research and development, five fail in product and market tests, and of the five that pass these tests, only two become commercial successes.[6]

If the expected returns from the introduction of the innovation do not exceed those obtainable from other investments by an amount that is large enough to justify the extra risks, the innovation should be rejected. If they do exceed those obtainable elsewhere by this amount, the profitability and risks involved in introducing the innovation at present must be compared with the profitability and risks involved in introducing it at various future dates. There are often considerable advantages in waiting, since improvements occur in the new products or process and more information becomes available regarding its performance and market. For example, in the case of new products, firms often employ test marketing to obtain additional information before making a full-scale commitment. (In test marketing, a sample of potential buyers is exposed to the product under more or less normal market conditions; from the results, the firm attempts to infer how some larger population of potential buyers will behave.)

There are disadvantages, as well as advantages, in waiting, perhaps the most important being that a competitor may beat the firm to the punch or that the conditions favoring the innovation may become less benign. In the case of new products, there is often a considerable disadvantage in not being first; sales opportunities will be lost in the interval that competitors are in the market ahead of this firm, and part of the market may be preempted. If the expected returns exceed those obtainable from other investments by an amount that is large enough to

[6] Booz, Allen and Hamilton, Inc., *Management of New Products,* New York, 1960.

justify the risks and if the disadvantages of waiting outweigh the advantages, the firm should introduce the innovation. Otherwise it should wait. Pioneering is a risky business; whether it pays off is often a matter of timing.

In recent years, a number of analytical devices have been developed to aid management in making decisions of this sort. Bayesian statistical techniques can be used to help decide whether or not to collect additional information before acting and if so, how much the additional information is worth. Such techniques can be useful in various ways. For example, they can help prevent market research from being carried out under circumstances where, regardless of the outcome of the research project, the optimal choice is unaffected or unclear; according to some marketing experts, such mistakes are common. Also, network techniques can sometimes help management to plan and schedule the activities leading up to the introduction of the new product or processes. For example, PERT can be used to identify critical schedule slippages and cost overruns in time for corrective action.

Assuming that the innovation is a new product and that it is successful, the innovator will have invested many times the original research costs leading to the basic invention by the time the new product is brought to market. According to the Department of Commerce's Panel on Invention and Innovation, the research and advanced development leading to the basic invention typically constitutes only about 5 to 10 percent of the total costs. The subsequent engineering and design of the product typically represent about 10 to 20 percent, while tooling and manufacturing engineering represent about 40 to 60 percent of the total costs. Finally, the manufacturing start-up expenses typically constitute about 5 to 15 percent and the marketing start-up expenses constitute about 10 to 25 percent of the total costs.[7]

[7] See U.S. Department of Commerce, *Technological Innovation,* Washington, January 1967. For a discussion of various analytical techniques, see E. Pessemier, *op. cit.* and the literature cited there.

## 4. Importance of External Sources

To what extent do new firms, firms in other industries, independent inventors, and universities play a leading role as innovators or as sources of the ideas underlying major innovations? Many economists emphasize the importance of these external sources of ideas and innovation, particularly in the less science-based industries. Recently, Arthur D. Little carried out an empirical study that sheds new light on the validity of this hypothesis.[8] The Little study is concerned with the origins of the recent major innovations occurring in three mature industries—textiles, machine tools, and construction. The results indicate that there have been relatively few innovations with major economic impact in these industries over the last twenty or thirty years. Those that have occurred have come primarily from outside the industry. To some extent, the ideas underlying them have come from independent inventors, from foreign technology, and from the formation of small new firms. But the most important source has been through the flow of technology from one industry to another. In some cases, these industries have borrowed technology from other industries; in other cases, another industry has entered their business, supplying new components, materials, or equipment; in still other cases, another industry has manufactured a new version of their product.

To illustrate the importance of external sources, let us take the case of numerically controlled machine tools, a very significant new development in metal-working technology. Numerical control provides instructions to machine tools in the form of coded instructions punched on paper tape and enables the machine tool itself to perform most of the functions done by the operator on conventional tools. Numerical control was not developed by the machine tool industry but by Massachusetts In-

[8] Arthur D. Little, "Patterns and Problems of Technical Innovation in American Industry," *The Role and Effect of Technology in the Nation's Economy*, Hearings before a Subcommittee of the Select Committee on Small Business, United States Senate, 88th Congress, First Session.

stitute of Technology which carried out the work for the U.S. Air Force. The first successful demonstration of a numerically controlled machine tool was held in the early fifties at M.I.T., and the first commercial version appeared at the Machine Show of 1955. It was used at first primarily by the aircraft industry on government projects. Although machine tool manufacturers like Giddings and Lewis played a significant role in its commercialization, the concept was developed outside the industry and much of the impetus behind the innovation came from the U.S. Air Force.[9]

The proportion of really major innovations developed and introduced by established firms is lower in these older industries than in technically sophisticated industries like chemicals, electronics, and aerospace. According to Donald Schon, the director of the Little study, the basic innovations in textiles over the past twenty years have come from the chemical industry, and the principal wave of technological change in the building industry has been a wholesale borrowing of methods from industrial manufacturing. In machine tools, we have seen that the concept leading to one of the most important innovations stemmed largely from external sources. In Schon's opinion, the older industries have difficulty in developing and introducing major innovations because they are fragmented into many small firms, the work of the industry is divided into many small, separately-controlled steps, they are committed to present methods and machines, and they spend little on research and development.[10] Finally, to put all of this in perspective, it should be noted that the many minor innovations, as well as the few major ones, play a very important role in determining the state and evolution of an industry. This point has, of course, been stressed before.

[9] See E. Schwartz and T. Prenting, "Automation in the Fabricating Industries," *Report to the President by the National Commission on Technology, Automation, and Economic Progress,* February 1966; and E. Mansfield, *Numerical Control: Diffusion and Impact in the Tool and Die Industry,* Small Business Administration, 1971. More will be said about numerical control in the final sections of this chapter.

[10] D. Schon, "Innovation by Invasion," *International Science and Technology,* March 1964. For further discussion and evaluation of these opinions, see Chapters II, III, and VI.

## 5. The Diffusion of New Techniques

As pointed out in Chapter II, technological change results in a change in the production function of an existing product or in an addition to the list of technically feasible products. In the American economy, firms and consumers are free to use new technology as slowly or as rapidly as they please, subject, of course, to the constraints imposed by the marketplace. How rapidly an innovation spreads is obviously of great importance; for example, in the case of a process innovation, it determines how rapidly productivity increases in response to the new process. In the remainder of this chapter, we discuss the nature of the diffusion process, the determinants of the rate of diffusion, and the characteristics of technological leaders and followers.

The diffusion process, like the earlier stages of the process of creating and assimilating new processes and products, is essentially a learning process. However, rather than being confined to a research laboratory or to a few firms, the learning takes place among a considerable number of users and producers. When the innovation first appears, potential users are uncertain of its nature and effectiveness, and they tend to view its purchase as an experiment. Sometimes considerable additional R and D is required before the innovation is successful; sometimes, despite attempts at redesign and improvement, the innovation never is a success. Information regarding the existence, characteristics, and availability of the innovation is disseminated by the producers through advertisements and salesmen; information regarding the reaction of users to the innovation tends to be disseminated informally and through the trade press.

Learning takes place among the producers of the innovation, as well as the users. Early versions of an innovation often have serious technological problems, and it takes time to work out these bugs. During the early stages of the diffusion process, the improvements in the new process or product may be almost as important as the new idea itself. For example, there were very important improvements in the catalytic cracking of petroleum in the period following the Sun Oil Company's first introduction of the process, the original Houdry fixed-bed process being outmoded in less than a decade. Moreover, when a new product's design is stabilized, costs of production generally fall in accord with the "learning curve." Thus, in the case of new machine tools, unit costs tend to be reduced by 20 percent for each doubling of cumulated output.[11]

Finally, the diffusion process involves the reallocation of resources. For example, if a new type of equipment is to replace an old type, labor, capital, and materials must be available to produce the new equipment, as well as the fuel and material it uses. The resources used to make the old equipment, and the fuel and material it uses, must be transferred to other employment. Some workers may be thrown out of work, at least temporarily; others must be retrained to operate the new equipment. Since physical capital, once constructed, tends to be relatively inflexible, changes in its allocation must occur largely through the building of new plant and equipment and the scrapping of old. The process of adjustment is often intricate and far-reaching, and the full exploitation of the social benefits from technological change requires that the process be as smooth as possible. Some of the problems of adjustment are discussed in Chapter V.

## 6. Rates of Diffusion

Once an innovation has been introduced by one firm, how rapidly does its use spread? Perhaps the earliest noteworthy study of this question was made in 1934.[12] The findings, based on data for twenty-three machines for periods ranging from eleven to thirty-nine years, indicate

[11] See J. Enos, *op. cit.*, and W. Hirsch, "Manufacturing Progress Functions," *Review of Economics and Statistics*, May 1956.

[12] H. Jerome, *Mechanization in Industry*, National Bureau of Economic Research, 1934.

> . . . the following estimates of the typical duration of periods in their life histories: commercial trial, three to eleven years; rapid increase in use, four to eleven years; slackened increase (with a customary annual gain of less than 10 percent), three to six years; decline, of undefined length. Processes and types of equipment suffer declines for long periods before they pass completely out of use. They linger on in small plants and for special uses long after they have been replaced by new processes or equipment in the major part of the industry.[13]

A more recent study [14] shows how rapidly the use of twelve innovations spread from enterprise to enterprise in four industries—bituminous coal, iron and steel, brewing, and railroads. The innovations are the shuttle car, trackless mobile loader, and continuous-mining machine (in bituminous coal); the by-product coke oven, continuous wide strip mill, and continuous annealing line for tin plate (in iron and steel); the pallet-loading machine, tin container, and high-speed bottle filler (in brewing); and the diesel locomotive, centralized traffic control, and car retarders (in railroads). Figure 4.1 shows the percentage of major firms that had introduced each of these innovations at various points in time. To avoid misunderstanding, three points should be noted in regard to these data. First, because of difficulties in obtaining information concerning smaller firms (and because the smaller firms often were unable to use the innovation in any event), only firms exceeding a certain size are included. Second, the percentage of firms having introduced an innovation, regardless of the scale on which they did so, is given. Third, in a given industry, most of the firms included in the case of one innovation are also included for the others. Thus the data for each of the innovations are quite comparable in this regard.

Two conclusions emerge from Figure 4.1. First, the diffusion of a new technique is generally a slow process. Measuring from the date of the first successful commercial application, it took twenty years or more for all the major firms to install cen-

[13] *Ibid*, pp. 20–21.

[14] E. Mansfield, *Industrial Research and Technological Innovation, op cit.*, Chapter VII.

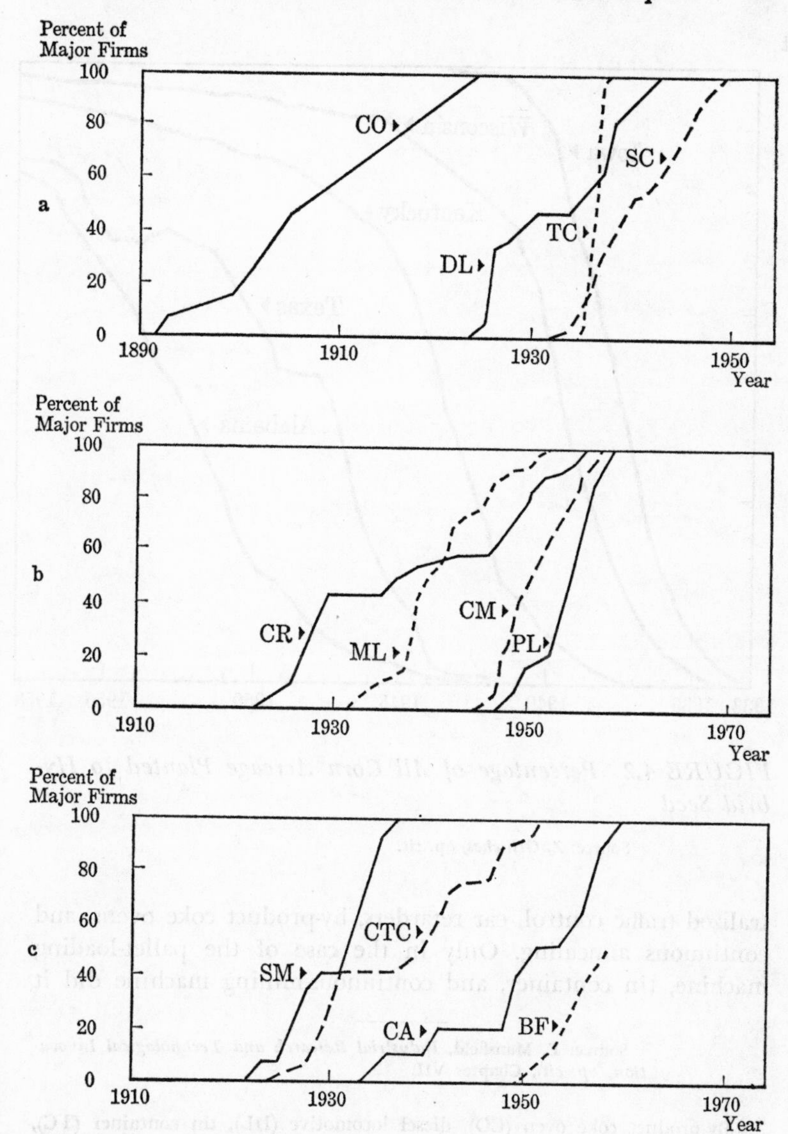


***FIGURE 4.1 Growth in the Percentage of Major Firms that Introduced Twelve Innovations, Bituminous Coal, Iron and Steel, Brewing, and Railroad Industries, 1890–1958***

**(Footnotes to Figure 4.1 are at the bottom of the next page.)**

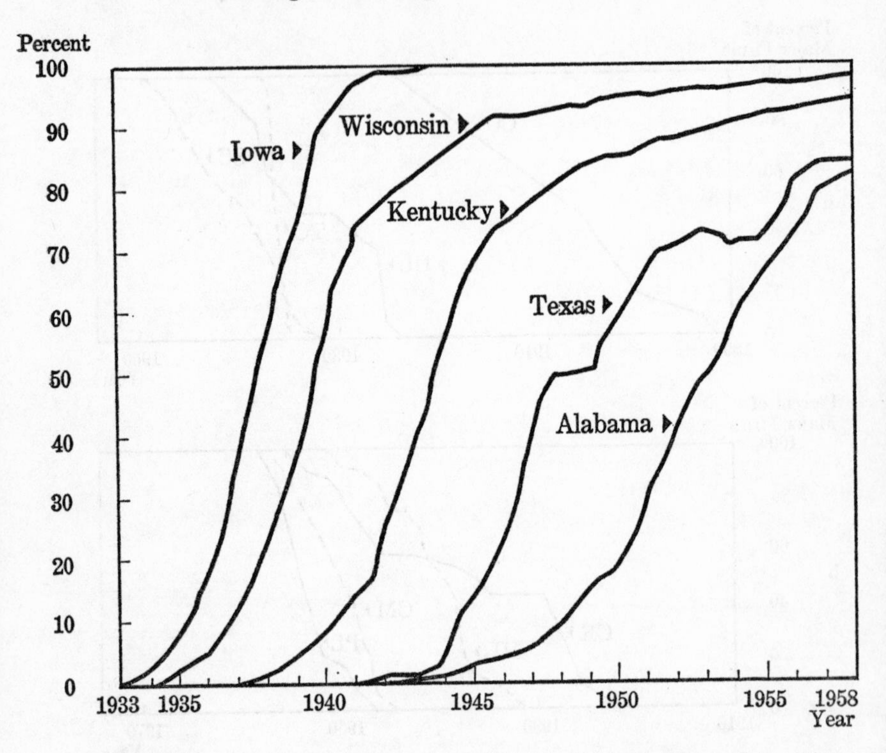


*FIGURE 4.2 Percentage of All Corn Acreage Planted to Hybrid Seed*

Source: *Z. Griliches,* op. cit.

tralized traffic control, car retarders, by-product coke ovens, and continuous annealing. Only in the case of the pallet-loading machine, tin container, and continuous-mining machine did it

---

Source: E. Mansfield, *Industrial Research and Technological Innovation, op. cit.,* Chapter VII.

[a] By-product coke oven (CO), diesel locomotive (DL), tin container (TC), and shuttle car (SC).

[b] Car retarder (CR), trackless mobile loader (ML), continuous-mining machine (CM), and pallet-loading machine (PL).

[c] Continuous wide strip mill (SM), centralized traffic control (CTC), continuous annealing (CA), and high-speed bottle filler (BF).

take ten years or less for all the major firms to install them. Second, the rate of imitation varies widely. Sometimes it took decades for firms to install a new technique, but in other cases they imitated the innovator very quickly. For example, fifteen years elapsed before half of the major pig-iron producers had used the by-product coke oven, only three years elapsed before half of the major coal producers had used the continuous-mining machine. The number of years elapsing before half the firms had introduced an innovation varied from 0.9 to 15.

In addition, studies have been made of the diffusion of hybrid corn, an important agricultural innovation. Griliches [15] made a valuable study of the differences among regions in the rate of diffusion of hybrid corn. Although serious research on hybrid corn was begun early in the century, the first application of the research results on a substantial commercial scale did not occur until the thirties. As shown in Figure 4.2, some regions began to use hybrid corn earlier than others; and once they had begun, some regions made the transition to full (or almost full) adoption more rapidly than others. For example, the time interval from 20 to 80 percent of full adoption required eight years in Alabama but only three years in Iowa. Ryan and Gross [16] confined their attention to about 250 farms in two small Iowa communities, some of their findings being as follows: First, the growth in the percentage of users followed an S-shaped curve. Second, the late adopters were not late because of lack of information concerning the existence of the innovation. By 1934, more than 90 percent of the farmers had heard of the new seed, but fewer than 20 percent had tried it. Third, the earliest adopters were very conservative in the percent of their acreage planted to hybrid corn during the year of their initial adoption.

[15] Z. Griliches, "Hybrid Corn: An Exploration in the Economics of Technological Change," *Econometrica,* October 1957. For a review of other studies of the diffusion process, see E. Rogers, *Diffusion of Innovation,* New York: The Free Press of Glencoe, 1962.

[16] B. Ryan and N. Gross, "The Diffusion of Hybrid Seed Corn in Two Iowa Communities," *Rural Sociology,* March 1943.

## 7. Determinants of the Rate of Diffusion

What determines an innovation's rate of diffusion? Before taking up this question, we should say a few words about the determinants of the ultimate, or equilibrium, level of use of the innovation. For a new process used to make an existing good or service, the equilibrium level of use depends upon the extent of its economic advantages over the other inputs it replaces, and on the sensitivity of the demand of the product it produces to any decline in price or increase in quality induced by the innovation. For a new final good, the equilibrium level of use depends on how much of this product consumers are willing to purchase at the price at which it can be produced and marketed profitably.

Four principal factors seem to govern how rapidly the innovation's level of utilization approaches this ultimate, or equilibrium, level: (1) the extent of the economic advantage of the innovation over older methods or products, (2) the extent of the uncertainty associated with using the innovation when it first appears, (3) the extent of the commitment required to try out the innovation, and (4) the rate of reduction of the initial uncertainty regarding the innovation's performance. Based on these factors, a simple mathematical model[17] has been constructed to explain the differences in the rate of diffusion shown in Figure 4.1. This model is based on the following four hypotheses:

First, as the number of firms in an industry adopting an innovation increases, it is assumed that the probability of its adoption by a nonuser increases. This assumption seems reasonable because, as experience and information regarding an in-

[17] E. Mansfield, *op. cit.*, Chapter VII. Also see E. Mansfield, "Technological Change: Measurement, Determinants, and Diffusion," *Report to the President by the National Commission on Technology, Automation, and Economic Progress*, February 1966. Note that this model pertains to the rate at which firms begin using the innovation. For the discussion to be complete, the intrafirm rate of diffusion must be considered too—and we shall do so in section 9.

novation accumulate, the risks associated with its introduction grow less, competitive pressures mount, and bandwagon effects increase.

Second, the expected profitability of an innovation is assumed to be directly related to the probability of its adoption. This seems reasonable because the more profitable the investment in an innovation promises to be, the greater will be the probability that a firm's estimate of its potential profitability will compensate for the risks involved in its installation.

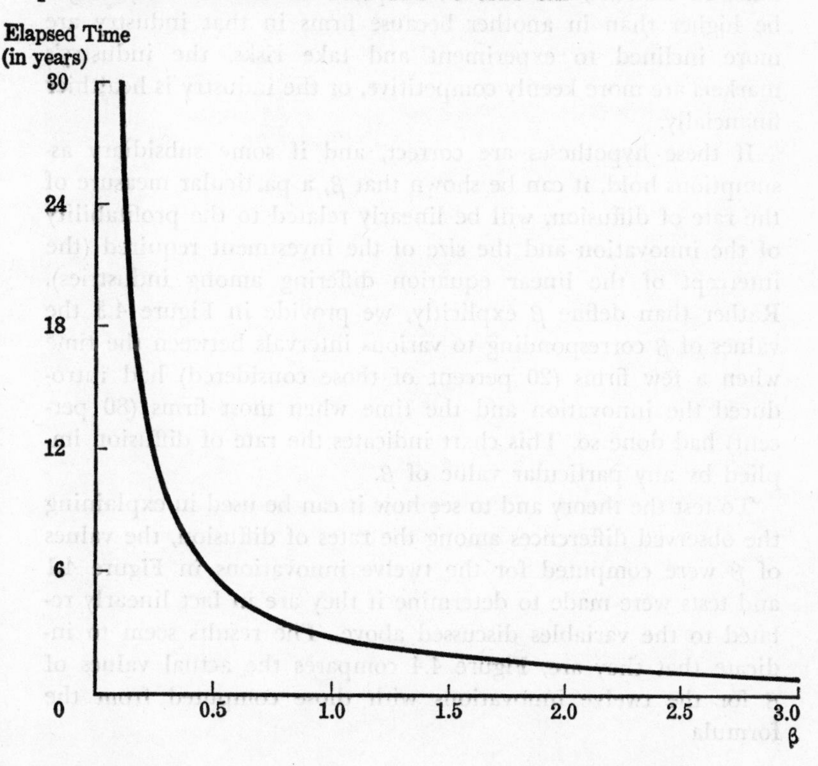


*FIGURE 4.3 Relationship between β*[a] *and the Number of Years Elapsed from the Time 20 Percent of Major Firms Had Introduced the Innovation to the Time when 80 Percent Had Done So*

[a] A particular measure of the rates of diffusion (see text).

Third, for equally profitable innovations, the probability of adoption is assumed to be smaller for innovations requiring relatively large investments. This is because firms will be more cautious before committing themselves to large, expensive projects, and they will have more difficulty in financing them.

Fourth, the probability of adoption of an innovation is assumed to be dependent on the industry in which the innovation is introduced. For equally profitable innovations requiring the same investment, the rate of adoption in one industry might be higher than in another because firms in that industry are more inclined to experiment and take risks, the industry's markets are more keenly competitive, or the industry is healthier financially.

If these hypotheses are correct, and if some subsidiary assumptions hold, it can be shown that $\beta$, a particular measure of the rate of diffusion, will be linearly related to the profitability of the innovation and the size of the investment required (the intercept of the linear equation differing among industries). Rather than define $\beta$ explicitly, we provide in Figure 4.3 the values of $\beta$ corresponding to various intervals between the time when a few firms (20 percent of those considered) had introduced the innovation and the time when most firms (80 percent) had done so. This chart indicates the rate of diffusion implied by any particular value of $\beta$.

To test the theory and to see how it can be used in explaining the observed differences among the rates of diffusion, the values of $\beta$ were computed for the twelve innovations in Figure 4.1 and tests were made to determine if they are in fact linearly related to the variables discussed above. The results seem to indicate that they are. Figure 4.4 compares the actual values of $\beta$ for the twelve innovations with those computed from the formula

$$\beta = \begin{Bmatrix} -0.57 \\ -0.52 \\ -0.29 \\ -0.59 \end{Bmatrix} + 0.53P - 0.027S,$$

in which the figures in the brackets pertain to the bituminous coal industry, the iron and steel industry, the brewing industry, and the railroad industry, respectively. $P$ is a measure of the relative profitability of an innovation, and $S$ is a measure of the size of the investment.[18]

It appears that this theory explains almost all the observed variation in the rates of diffusion. As shown in Figure 4.4, the theoretical relationship between $\beta$ on the one hand, and $P$ and $S$ on the other, seems to hold very well. Although the model is still in the experimental stages, it seems to be a promising forecasting device. In addition, two other points should be noted: First, there is an apparent tendency for the rate of diffusion to be higher when the innovation does not replace very durable equipment and when an industry's output is growing rapidly. However, neither of these apparent tendencies is statistically significant. Second, although there is too little data to warrant a conclusive statement the interindustry differences seem roughly consistent with the hypothesis that rates of diffusion are higher in less concentrated industries.

Of course, this simple theory cannot include all of the important factors influencing the rate of diffusion. The frequency and extent of advertising and other promotional devices used by producers will also have an influence. So will the innovation's requirements with regard to knowledge and coordination, the diffusion process being delayed if the innovation requires new kinds of knowledge on the part of the user, new types of behavior, and the coordinated efforts of a number of organizations. In addition, the diffusion process may be impeded by bottlenecks in the production of the innovation; for example, it took years before all the early orders for the Boeing 707 could be filled. Also, if an innovation requires few changes in sociocultural values and behavior patterns, it is more likely to spread rapidly. And the more apparent the profitability of an inno-

[18] $P =$ average pay-out period to justify investments (during the relevant period) divided by average pay-out period for investment in the innovation. $S =$ average initial investment in the innovation divided by average total assets (from *Moody's*) for the relevant period. For a more detailed description of P and S, *ibid.* The figures in the formula are least-square estimates.

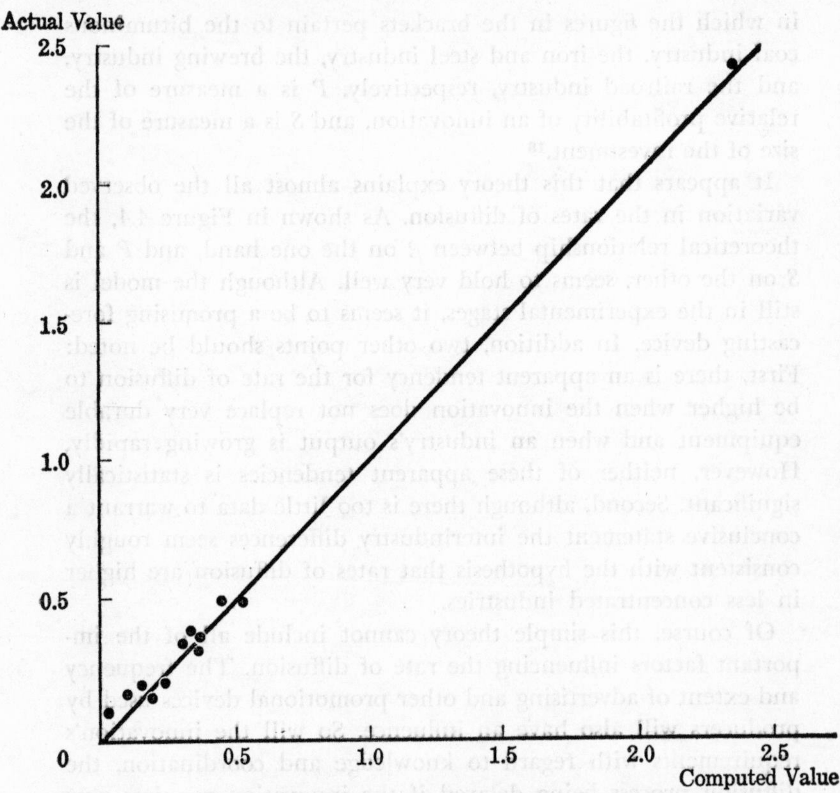


Source: *E. Mansfield,* Industrial Research and Technological Innovation, op. cit., *Chapter VII.*

*FIGURE 4.4 Comparison of Actual and Computed*[a] *Values of β for Twelve Innovations*

[a] Computed from the equation given in the text. Difference in values is measured by the vertical distance from the circle to the line; the line represents perfect correspondence.

vation and the easier it is to explain and demonstrate the advantages of the innovation, the more quickly it is likely to spread. In addition, the policies adopted by relevant labor unions can have an important influence on the rate of diffusion, as we shall see in Chapter V.

## 8. Technical Leaders and Followers

What are the characteristics of firms that are relatively quick—or relatively slow—to begin using new techniques?[19] Seven characteristics of a firm and its operations might be expected to affect a firm's speed of response to a new technique: (1) size of firm, (2) expectation of profit from the new technique, (3) rate of growth of the firm, (4) the firm's profit level, (5) the age of the firm's management personnel, (6) liquidity of the firm, and (7) the firm's profit trend. Let us consider these characteristics in order.

SIZE OF FIRM. One would expect larger firms to introduce a new technique quicker than small ones. In industries where the larger firms are most progressive, this would certainly be expected.[20] But even if the larger firms are not relatively progressive and do not introduce more than their share of the innovations, one would expect them to be quicker, on the average, to begin using new techniques than smaller firms. To illustrate this, we assume an industry with two firms, one large (80 percent of the market), one small (20 percent of the market). If the large firm does its share of the innovating, it will be first in 80 percent of the cases—and it will be quicker on the average than the small firm.[21] Empirical studies substantiate the hypoth-

[19] The results in this section are based on E. Mansfield, *op. cit.*, Chapter VII. They are based on data pertaining to the steel, coal, railroad, and brewing industries.

[20] The advantages of the large firm are fairly obvious—greater financial resources, bigger engineering departments, better experimental facilities, closer ties with equipment manufacturers, and so forth.

[21] We have shown elsewhere that if firms differ only in the number of machines, the probability distribution of the lag before adoption of a new technique being the same for each machine, the larger firms will be quicker, on the average, than the smaller firms to begin using the new technique. See E. Mansfield, *op. cit.*, Chapter VIII. Even if the largest firms in some industries do less than their share of the innovating, there is likely to be an inverse relationship, on the average, between size of firm and how long a firm waits before it begins to use a new technique.

esis that large firms are quicker, on the average, than small ones to begin using new techniques. Moreover, they provide estimates of the extent of this effect; in the coal, steel, railroad, and brewing industries, the elasticity of delay with respect to firm size is about —0.4.

EXPECTED PROFIT FROM THE NEW TECHNIQUE. The higher the expected return from the new technique, the quicker it would be expected to be adopted. Introduction of a new technique would be delayed if the return was not deemed adequate to offset the risk involved. Unfortunately, only partial data can be obtained regarding firms' profit expectations with respect to various new techniques. However, on the basis of available data, there are strong indications that the profitability of the investment is important in determining how rapidly a firm adopts a new technique.

GROWTH RATE OF THE FIRM. The more rapidly a firm is growing, the more responsive it could be expected to be in adopting a new technique. An expanding firm can introduce a new technique into its new plants, whereas a firm that is not growing has to wait until it can profitably replace existing equipment. Several studies suggest that this factor may often be important; but our own results revealed no close relationship between a firm's rate of growth and the rate at which it adopts a new technique. The effect of the factor was not statistically significant.

THE FIRM'S PROFIT LEVEL. More prosperous firms might be expected to adopt a new technique more quickly than firms with low levels of profit. Since less prosperous firms have smaller cash inflows and poorer credit ratings, they experience greater difficulty in financing the investment. They are, therefore, in a less favorable position to take the risk involved in being one of the first to adopt a new technique. Despite these factors, however, the available data show no close relationship between a firm's profit rate and the rate at which it adopts a new technique. The effect of this factor was not statistically significant.

THE AGE OF MANAGEMENT PERSONNEL. Firms with younger top management personnel might be expected to adopt a new technique more quickly than those with older top managements. It is often asserted that younger managements are less bound by traditional ways, and some evidence does indicate that this is true in agricultural enterprises. However, for industrial firms there is no evidence that this is the case, the effect of the factor being statistically nonsignificant.

LIQUIDITY OF THE FIRM. The more liquid firms might be expected to begin using a new technique more quickly than the less liquid firms, because the former are better able to finance the investment. The testing of this hypothesis was limited to only a few cases because of the lack of sufficient data. In these cases, the effect of liquidity on a firm's speed of response was not statistically significant.

THE FIRM'S PROFIT TREND. Firms with decreasing profits might be expected to be more responsive to a new technique, because they might be expected to search more diligently for new alternatives. This hypothesis could be tested for only a few cases because of insufficient data, and in these cases the effects of a firm's profit trend were not statistically significant.

## 9. Intrafirm Rates of Diffusion of a New Technique

Sections 7 and 8 were concerned with the rate at which firms begin to use new techniques. This section examines the influence of various factors on the intrafirm rate of diffusion, that is, the rate at which a firm, once it has begun to use a new technique, continues to substitute it for older methods. We report the findings of a study of the diffusion of the diesel locomotive. Table 4.3 shows, for thirty randomly chosen railroads, the number of years that elapsed between the time when diesel loco-

*TABLE 4.3 Time Interval between Date When Diesel Locomotives Were 10 Percent of All Locomotives and Date When They Were 90 Percent of All Locomotives in Thirty Randomly Chosen Class I Railroads*

| *Time Interval* | *Number of Firms* |
|---|---|
| 14 or more years | 3 |
| 11–13 years | 7 |
| 8–10 years | 11 |
| 5–7 years | 3 |
| 3–4 years | 6 |
| Total | 30 |

Source: *E. Mansfield,* Industrial Research and Technological Innovation, op. cit., *Chapter IX.*

motives were 10 percent of the total locomotive stock and the time when they were 90 percent.[22] There is wide variation among firms in the intrafirm rate of diffusion. Although an average of nine years was required for the thirty firms to increase dieselization from 10 to 90 percent of complete conversion, six firms took only three or four years to convert and three took fourteen years or more.

The results of an econometric study show that about two-thirds of the variation in the rate of intrafirm dieselization among the railroads can be explained by the following factors: profit expectation of the investment in diesel locomotives, the date when a firm began to dieselize, size of the firm, the age distribution of its steam locomotives, and a firm's initial liquidity. Several points should be noted regarding these findings. First, together with previous results, the findings suggest that there exists an important economic analogue to the classic psychological laws relating reaction time to the intensity of the stimulus. The profitability of an investment opportunity acts as a stimulus, the intensity of which seems to govern quite closely a firm's speed of response. Second, small firms, once they began, were quicker than their larger rivals to substitute the

[22] We present results based on utilization data, as well as these ownership data, in *ibid.,* Chapter IX.

new technique for the old. Third, the results point up the importance of when a firm begins to use the new technique, the age of its equipment at that time, and its liquidity. All of these factors have a statistically significant effect on the intrafirm rate of diffusion, (measured in terms of either ownership or utilization data or both).[23] However, as so often has been the case in studies of investment behavior, the effect of the profitability of the firm is not statistically significant.

## 10. Numerical Control: A Case Study

In the final sections of this chapter we present a brief case study describing the diffusion of one of the most important new techniques that has arisen in this century—numerical control. Numerical control of machine tools is a way of operating them by means of numerical instructions expressed in code. Prepared in advance and recorded on tape or cards, these instructions control the sequence of machine operations. Numerical control results in a great many economic advantages—lower labor costs, lower capital costs, lower inspection costs, lower inventory costs, and higher quality of output. According to one leading research institute, numerical control "is the most significant new development in manufacturing technology since Henry Ford introduced the concept of the assembly line".[24]

In section 4, we pointed out that numerical control was not

[23] As one would expect, the intrafirm rate of diffusion was faster for firms that were later to begin using the innovation, that had older equipment, and that were more liquid. The age of a firm's old equipment and size of firm have a statistically significant effect only when the utilization rather than the ownership data are used.

[24] Illinois Institute of Technology Research Institute, *Technological Change: Its Impact on Metropolitan Chicago,* 1964, p. 1. For a more detailed discussion of the topics taken up in this and the following two sections, see Edwin Mansfield, *Numerical Control: Diffusion and Impact in the Tool and Die Industry,* Small Business Administration, 1971; and references cited there. For a summary of some of the results, see E. Mansfield, J. Rapoport, J. Schnee, S. Wagner, and M. Hamburger, *Research and Innovation in the Modern Corporation,* W. W. Norton, forthcoming, Chapter 9.

developed by the machine tool builders but by sources outside the industry. The research and development leading to numerical control seems to have gotten started in a serious way in about 1947. It was stimulated by the need for new machining methods to produce the many intricate parts of modern aircraft quicker, cheaper, and more accurately than conventional methods. John T. Parsons, who was the owner of a small firm in Michigan that produced rotor blades, won a study contract from the Air Force after conceiving a jig borer that was coupled with automatic data processing equipment. Parsons turned for assistance to the Servomechanism Laboratory of the Massachusetts Institute of Technology. Once the idea was shown to be feasible, MIT was given a direct contract for the development of an experimental milling machine. By the fall of 1952, MIT scientists and engineers had successfully developed the first such milling machine. Refinements were made at MIT and by numerous machine tool builders and by producers of control and computer equipment. In 1955, the first few commercial models were displayed at the National Machine Tool Show and placed in plants for operational use.

Numerically controlled machine tools are more expensive than conventional tools. That is, the first cost of a numerically controlled machine tool is relatively high. For example, with very few exceptions, the installed cost of a numerically controlled milling machine is $50,000 or more. Until about 1960, numerically controlled machine tools were so expensive and experimental that they were bought largely for defense and related operations (particularly in the aerospace industry), the government footing the bill. A major task confronting the machine tool manufacturers was to devise a practical, workable machine that could be priced attractively for commercial customers, a machine that little shops as well as big ones could afford. In 1961, Pratt and Whitney came out with such a machine, its Tape-O-Matic numerically controlled drilling machine, about 1,000 of which had been sold by 1964. Other tool builders soon put competing low-priced numerically controlled drilling machines on the market.

## 11. The Diffusion of Numerical Control

Since 1955, the use of numerical control by American industry has grown rapidly. In 1959, the first year for which annual figures are available, numerically controlled machine tools constituted 3.7 percent of the value of shipments of all metal-cutting machine tools. This percentage increased to about 12 percent in 1964 and to about 20 percent in 1966. Among the major users of numerically controlled machine tools are the aircraft industry and the metal-working machinery industry. In 1954–1963, these two industries accounted for shipments of about one-third of such tools.

To illustrate the rate of diffusion of this innovation, consider the important tool and die industry. This industry is composed of thousands of small firms, none of which has anything approaching monopoly power. According to the best estimates available, only about 1 percent of the firms in the National Tool, Die, and Precision Machining Association had begun using numerical control by the beginning of 1961. By the beginning of 1966, the percentage had grown to 10 percent; and by the beginning of 1968, 20 percent of the firms had begun using numerical control.

What sorts of firms have been relatively quick to begin using it? On a priori grounds, there are a great many reasons for expecting the larger tool and die firms to be quicker, on the average, than the smaller ones to introduce numerical control. For example, the larger firms are more likely to have the financial resources to enable them to experiment, and they are more likely to have the technical know-how and the managerial qualities that are so important in determining a firm's speed of response to a new technique. This hypothesis is borne out by the facts. The median employment of firms using numerical control at the beginning of 1968 was about 60, while the median employment of non-users was about 30. Moreover, among the users of numerical control, there is a significant inverse relationship between the size of the firm and the year when the firm began using numerical control.

Another variable that would be expected to influence whether or not a tool and die firm adopted numerical control relatively quickly is the education of the firm's president. Better educated entrepreneurs are likely to be in a better position to understand the issues regarding numerical control, to have the flexibility of mind to use it, and to be in contact with technical and university centers and the relevant literature. The data seem to bear out this hypothesis, most of the users (for which we have data) being college graduates but most of the non-users having finished only high school or less. The difference is statistically significant.

Still another variable that would be expected to influence whether or not a firm adopted numerical control is the age of the firm's president. Younger entrepreneurs would be more likely to make the break with the past, their emotional attachment to old skills and old technology being weaker and their willingness to take risks probably being greater than their older rivals. The available data are consistent with this hypothesis, the median age of the users being about 48 and the median age of the non-users being about 55. However, age and education are correlated between themselves, and when both variables are included in the analysis, the effect of education is statistically significant, but the effect of age is not.

## 12. Numerical Control in the Ford Motor Company

Numerical control has a promising role in the automobile industry. For example, consider the case of the Ford Motor Company. During the fifties, it was believed that automotive body tooling, because of its complexities, would be a very promising area for numerical control. Construction of templates, wood models, dies, and related body tooling are long, complex operations that have been partly responsible for the long delay between a clay model and the marketing of a new car. Because car body surfaces are created by persons concerned primarily with aesthetic features, it turned out that computer programs designed for analytical surfaces in the aerospace industry could not be used. The problem required original solutions.

In 1958, Ford decided that the potential gains warranted undertaking a program aimed at overcoming the disadvantages of conventional methods. Systems were developed and experimental equipment was designed or procured. When the basic principles were proved out, a "Numerical Control Activity" was organized, its responsibility being to develop applicable systems, computer programs, processing, and shop methods. Engineers, mathematicians, programmers, and technicians were progressively assigned to this activity. A major effort was made by this group over a period of several years, in cooperation with other groups such as manufacturing engineering, product engineering, engineering research, and the operating divisions.

Basic mathematical solutions of three-dimensional problems were written and converted into computer programs. Commercial equipment was supplemented where necessary with special devices. Gradually, the system's capabilities were increased to a level that would insure the production of quality surfaces in day-to-day operations. Additional features transcending computation of car surfaces had to be provided for die requirements. Experienced die processors and die design personnel worked with mathematicians and computer programmers in defining each requirement. In other areas, numerical control was adopted to machine flat and geometrically defined surfaces.

How has the program worked out? Ford engineers and management seem very enthusiastic about the results. Die surface machining can be started several weeks earlier in a new model program, and excellent surface continuity and accuracy are obtained. According to the manager of Ford's Numerical Systems Department, "dies can be produced accurately and faster by numerical control than by conventional methods. Therefore, dies now requiring long lead-time, or particularly critical dies, will be made by numerical control. It is anticipated that our die suppliers will gradually develop pertinent NC techniques and personnel, and become self-sufficient in NC die machining."[25]

In conclusion, numerical control is only one of many new developments that are shaping the nature and evolution of our economy in the seventies. Each of these developments comes about in a different way and has different effects. No single, brief case study can adequately represent all of these developments. Nonetheless, this example should provide insight into the process of technological change. It provides information concerning one of the most important developments in manufacturing, a development that will have marked effects on the economy in the last third of the twentieth century.

[25] N. Hopwood, "Numerical Control at Ford Motor Company," *Proceedings of the Third Annual Meeting of the Numerical Control Society*, April 1966, pp. 24–25.

CHAPTER FIVE

# Automation, Labor, and Government

## 1. Technological Unemployment

When technological change is mentioned, the first thing many people think of is unemployment. The fear of technological unemployment is by no means new. During the mid-1700's, a mob of worried English spinners smashed into James Hargreave's mill, and destroyed the first workable multi-spindle frames. Similar forms of labor resistance to the adoption of new techniques are chronicled in the histories of most major nations. Moreover, resistance of this kind is not confined to unskilled workers. Although new techniques often result in the displacement of unskilled workers, there have been many cases where skilled journeymen were as much affected as unskilled. For example, the Owens automatic glass-blowing machines largely destroyed the bottle-glassblowers' craft.

Early social attitudes contributed to the fear of and resistance to new techniques by many worker groups. In the late 1800's and the early part of this century, the prevailing attitude was "sink or swim." Little or no help was available to displaced workers. If a worker's skills were outmoded, he was often forced into jobs (when they were available) which were far below his previous status. In accord with the mores of the times, it was felt that any efforts to help would be quite wrong; each man should stand on his own two feet. In recent years, attitudes have changed. There is a growing feeling that society, which is, by and large, a beneficiary of technological change, has an obligation to minimize the losses and assist the readjustment of those who are hurt. In this chapter, we begin by discussing the relationship between technological change and aggregate unemployment, and the attitudes of unions toward management's adoption of new techniques.

## 2. Automation

Although the dread of technologically induced mass unemployment is not new, it was revived with great effect in the late fifties and early sixties, the new scare word being "automation." For example, Professor Crossman of Oxford University, addressing an international conference in 1964, said that "unemployment due to automation will grow steadily over the next few decades, perhaps centuries, and in the end it is likely to reach a very high figure, say 90 percent of the labor force, unless radical changes are made in the present pattern of working."[1] Although "automation" means different things to different people, it generally refers to processes designed to mechanize the human cognitive, conceptual, and informational processes. First, there are automatic control mechanisms, which introduce the closed-looped feedback principle and make possible the creation of an automatic remote-controlled, self-contained production system. "Feedback" is a concept of control by which the input of a machine is regulated by the machine's own output, the consequence being

[1] Organization for Economic Cooperation and Development, *The Requirements of Automated Jobs,* Paris, 1965, p. 21.

that the output meets the conditions of a predetermined objective—as in a thermostatically controlled heating system. Process control machines have found use in oil refineries and chemical plants, as well as many other industries. Second, there are transfer machines, commonly called "Detroit automation," which have been employed in the automobile industry. Such equipment has been used to machine cylinder blocks. Third, there are a variety of uses of computer technology. Although we have described some of these uses in Chapter II, it may be worthwhile citing others. For example, consider American Airlines' reservation system. A central computer is connected to small desk machines displaying flight information. By pushing various buttons, a ticket agent can report a sale (or concellation) or request information. The central computer flashes back replies to inquiries on space availability. When a sale or cancellation is made, the correct number of seats is added to, or deducted from, the inventory of seats in the computer's memory. Thus, sales are not approved if space is not available, and expensive time lags are eliminated. It requires only a tenth of a second to tell if a seat is available on any flight leaving New York.[2]

Another example is furnished by Westinghouse Electric Corporation's Tele-Computer Center. Westinghouse, which probably uses computers as widely as any American firm, uses this center to provide up-to-date information regarding costs, sales, and other critical variables and to simulate various aspects of its business. In addition, the center handles about 2,000 orders a day for various products, prepares invoices, and does the bookkeeping; sends incoming orders to the nearest warehouse with the product in stock; and automatically adjusts warehouse stocks to optimum level by sending a reorder to the factory. It has enabled the firm to close six of its twenty-six warehouses, to cut inventories by 35 percent, and to provide better service.[3]

[2] See W. Buckingham, "Gains and Costs of Technological Change," in G. Somers, E. Cushman, and N. Weinberg, *op. cit.*

[3] See *Business Week,* June 25, 1966; and E. Mansfield, *Managerial Economics and Operations Research,* New York: W. W. Norton, revised edition, 1970.

## 3. Technological Change, Aggregate Demand, and Structural Unemployment

Is it true that increases in the rate of technological change necessarily result in increases in aggregate unemployment? Contrary to much popular opinion, the answer is no. Changes in aggregate unemployment are governed by the growth in the aggregate demand for goods and services and the growth in the labor force, as well as the growth in output per man-hour.[4] If the rate of increase of aggregate demand equals the rate of increase of productivity plus the rate of increase of the labor force, there will be no increase in aggregate unemployment, regardless of how high the rate of increase of productivity may be. Although there will be increases in some types of jobs and decreases in others, the total number of unemployed will not be affected.

Thus, rapid technological change need not result in increased aggregate unemployment. The important thing is that the government increase aggregate demand at the proper rate. If aggregate demand increases too slowly, increases in aggregate unemployment will take place. If aggregate demand increases too rapidly and resources are already fully employed, inflation will result. Unfortunately, there is nothing that insures that aggregate demand will grow at the right pace—as witnessed by the fact that it grew too slowly in the thirties and too rapidly immediately after World War II. However, through appropriate fiscal and monetary policies, the Federal government can compensate for inadequate or too rapid rates of growth of aggregate demand. The job of choosing and carrying out appropriate fiscal and monetary policies is not always easy, but the problems are by no means insoluble.

During the fifties and early sixties, there was considerable concern that workers and jobs were becoming more and more mismatched. According to the "structuralists," new methods and

[4] Changes in the average hours of work also influence aggregate unemployment, but their effect has been quantitatively less important in recent years.

equipment were increasing the skill and educational requirements of available jobs, and making it more likely that shortages of highly educated workers would coexist with unemployed pools of unskilled workers. Other economists denied that there was a substantial increase in the amount of structural unemployment, that is, unemployment that exists because the workers available for employment do not possess the qualities that employers with unfilled vacancies require. An important and lively debate took place, both inside and outside the government.

The important question was how unemployment would respond to an increase in the general level of demand. If an increase in demand would result in a reduction of unemployment, then evidently at least that much unemployment was not structural; if it would fail to reduce unemployment, then the unemployment that remained could be called structural. To put the structuralist hypothesis to a test, attempts were made to compare various periods when the general pressure of demand was about the same, the purpose being to see whether the level of unemployment was higher or more strongly concentrated in certain skill categories, industries, or regions in more recent periods than in the past.

The results, as well as the course of events since the 1964 tax cut, provided little support for the structuralist view. There was no evidence that unemployment was becoming more concentrated in particular geographical regions. Moreover, after adjusting for the effect of the over-all unemployment rate on the unemployment rate in particular occupations and industries, there were very few occupations or industries showing a statistically significant tendency for unemployment to increase with time. Also, after adjusting for the effect of the over-all unemployment rate, there was no significant tendency for unemployment rates among Negroes to rise during the late fifties and early sixties. Although a substantial portion of all unemployment may have been structural, there has been no evidence that unemployment of this type increased greatly in recent years.

## 4. Labor Displacement

Although an adequate level of aggregate demand can go a long way toward assuring that aggregate unemployment will not exceed a socially acceptable minimum, it cannot prevent labor from being displaced from particular occupations, industries, and regions, and being drawn to others. Nor would we want to eliminate such movements of labor, without which it would be impossible to adjust to changes in technology, population, and consumer tastes. However, regardless of the long-run benefits of this adjustment process, important problems may arise in the short run, great distress being imposed on the workers who are displaced. It is important that these movements of labor be carried out as efficiently and painlessly as possible.

The most serious adjustment problems have occurred when massive displacement has occurred in isolated areas among workers with specialized skills and without alternative sources of employment. Coal miners are a good example. About two-thirds of the nation's bituminous coal miners in 1960 were located in West Virginia, Pennsylvania, and Kentucky, and most of the coal mining was concentrated in isolated towns. Employment in bituminous coal started to decline after World War II, due partly to shifting demand and to the adoption of new techniques. Between 1947 and 1959, total employment in the industry declined by more than 60 percent, the consequence being that about 10 percent of the labor force in five major bituminous coal areas were unemployed during most of the fifties. In twenty-five smaller areas where coal mining was of more importance, the unemployment rate was even higher.[5]

The biggest of all displacements has been in agriculture, where the number of farm owners and farm workers declined by over 40 percent during the postwar period. Modern farm techniques —ranging from chemical fertilizers and insecticides to the cotton

[5] S. Levitan and H. Sheppard, "Technological Change and the Community," in G. Somers, E. Cushman, and N. Weinberg, *Adjusting to Technological Change,* New York: Harper and Row, 1963.

picker and huge harvesting combines—have contributed to this exodus, although they are by no means the only reason. Of those who left agriculture, many "suffering from deficient rural educations, lacking skills in demand in urban areas, unaccustomed to urban ways, and often burdened by racial discrimination, exchanged rural poverty for an urban ghetto." [6]

Displaced older workers have encountered particularly serious problems. Seniority rights have functioned to hold down the displacement of older workers; but once unemployed, they are less likely than younger workers to be re-employed. For example, one year after the shutdown of the Packard plant in Detroit, 77 percent of the under-forty-five-year-old workers were working at another job, whereas the percentages were 67 for the forty-five to fifty-four age group and 62 for the fifty-five to sixty-four age group. Even if skill is held constant, the older workers had more serious unemployment problems than the younger workers, at least as measured by length of unemployment. Another study obtained similar results for a period when the general level of employment was higher than in the Packard case.[7]

It is easy to understand why older workers have more difficulties in finding another job. Many of their skills are specific to a particular job, and much of their income and status may be due to seniority, the consequence being that they cannot command as high a wage on the open market as they previously earned. Moreover, employers naturally are reluctant to invest in hiring and training a worker who will only be available for a relatively few years and who often has relatively limited education. Finally, because of their roots in the community, older workers are less likely to move to other areas where jobs are more plentiful.[8]

Displaced Negroes seem to find it even more difficult than

[6] *Report of the National Commission on Technology, Automation, and Economic Progress, op. cit.*, p. 20.

[7] See H. Sheppard, L. Ferman, and S. Taber, *Too Old to Work—Too Young to Retire: A Case Study of Permanent Plant Shutdown*, Senate Special Committee on Unemployment Problems, 1959.

[8] See A. Ross and J. Ross, "Employment Problems of Older Workers," *Studies in Unemployment*, Senate Special Committee on Unemployment Problems, 1959.

older workers to obtain suitable re-employment at their previous status and earnings. For example, among workers who had been earning the same wage at Packard, Negroes experienced a higher average length of unemployment than whites. Moreover, the prestige and economic level of the new job were more likely to be lower for Negroes than whites.[9] Needless to say, prejudice often plays an important role in preventing displaced Negroes from obtaining new jobs.

## 5. Union Policies Toward New Techniques

Union policies toward management's adoption of new methods and equipment can be classified into five types. First, there is willing acceptance, which is the most frequent policy. This policy is adopted in the numerous cases where the new technique makes little difference with respect to skill requirements and number of jobs, but where the productivity gains make it attractive to labor by providing greater opportunity to bargain for wage increases. Moreover, unions may be led by the bargaining process or by the nature of the economic situation to accept willingly new techniques that involve a mixture of advantages and disadvantages. For example, the United Mine Workers gave the employers a free hand in making such changes, despite the fact that the industry had not been an expanding one. John L. Lewis, the head of the union until 1960, preferred to allow the employers to raise productivity and then compel them to pass on much of the gains to the workers in the form of higher wages.

Second, there is outright opposition to change. Although this policy is not rare, it is adopted in only a small proportion of cases, because unions know that it is unlikely to succeed for more than a short period and because bargaining is likely to be more advantageous than uncompromising opposition. When it occurs, opposition takes various forms. In some cases, the union

[9] See J. Hope, "The Problem of Unemployment as it Relates to Negroes," *Studies in Unemployment, ibid.*

has refused to use the new technique; for example, some locals of the painters' union have refused to use the spray gun. In other cases, the union has refused to use the new equipment efficiently; for example, the compositors' union has restricted the use of the teletypesetter by requiring that the machine be operated by a journeyman printer or apprentice. In still other cases, where the use of the new technique requires new rates or changes in seniority rules, the union may block the change by refusing to negotiate on them.

Third, there is competition with the new technique. According to this policy, which is a form of opposition, the union tries to compete with the innovation either by encouraging the more efficient use of an old method or the use of an alternative one. This policy is adopted when there are some types of work for which the new technique has little advantage or when the new technique is less costly but the old process produces better quality. For example, the lathers' and plasterers' unions, in cooperation with their contractors, have advertised the alleged deficiencies of dry wall.

Fourth, there is encouragement of technological change and the adoption of new techniques. The difference between this policy and a policy of acceptance is that in this case the union plays an active role in promoting change. This policy is usually pursued when the union is worried about the competitive position of an industry or a plant. For example, in the needle trades, profit margins tend to be small and mortality rates of firms tend to be high. The International Ladies' Garment Workers Union and the Amalgamated Clothing Workers have engineering departments; and although it is not their principal purpose to help employers adopt better techniques, they do provide such help in cases where the union has special reason to assist an employer.

Fifth, there is adjustment to change. The essence of this policy is an effort by the union to control the use of the new technique and to deal with the opportunities and problems it presents. Since no two innovations are alike, carrying out a policy of adjustment means negotiating tailor-made agreements on a wide assortment of issues. For example, who will do the work on the new equipment? A new technique may alter the kind of skill needed, it may transfer work from one seniority district to another, or it may greatly reduce the degree of skill required. Such changes often pose extremely difficult problems for the union. Jurisdictional disputes may occur with other unions, and internal conflicts may occur within a particular union.

What will be the rate of pay? If the new method increases or maintains the degree of skill required, the union may attempt to negotiate a wage increase. For example, the pulp and sulphite workers found that new methods usually did not reduce the number of workers but often increased skill requirements. Thus the union concentrated on trying to make these changes yield wage increases. If the new method reduces the skill requirements, the union may try hard to get an agreement that men operating the new equipment will receive the same rates as workers on old processes. When a new method is introduced that obviously means large savings to the company, it is difficult, and often foolish, for management to refuse to share the gains.

If workers are displaced, what is to be done for them? The union may want to increase the number of jobs on the new process, thus making work for some or all of these workers. (Make-work policies of this sort are discussed more fully below.) If the amount of displacement is small, the major concern of the union may be to find them other jobs and to obtain severance pay for them. For example, there was a strike in 1958 against Columbia Broadcasting System over job security for the men expected to be displaced by the introduction of video tape. The strike was settled when provision was made for severance pay of up to thirteen weeks' wages for layoff or dismissal due to automated processes.[10]

## 6. Determinants of Union Policy

The policy adopted by a union depends primarily on three factors. First, it depends on the nature of the union. A policy of

[10] See S. Slichter, J. Healy, and R. Livernash, *The Impact of Collective Bargaining on Management*, Washington, D.C.: The Brookings Institution, 1960.

opposition, when it occurs, is usually adopted by craft unions. Industrial unions usually find that the adverse effect of a new technique is limited to a small portion of their membership and that, whereas some members are hurt, others are better off. Consequently, they are likely to pursue a policy of adjustment. Moreover, industrial unions are usually less interested than craft unions in who will hold the jobs on the new process. Since the industrial union covers all workers in the plant, the workers on the new process remain within the bargaining unit; but a craft union faces the possibility that the work on the new process will be outside of its jurisdiction.

Second, the union's policy depends on the nature of the industry, firm, or occupation. If the employers face serious competition and if employment would be increased if better methods were used, the union may encourage the adoption of new techniques. If the industry or firm is experiencing rapid growth, the union may try to negotiate the best possible rates on the new jobs, whereas the union's policy is less predictable if the industry or firm is contracting. If the union believes that new techniques will have no effect on the rate of contraction, it may try to limit displacements through make-work rules. This, for example, has been the reaction of some of the railroad unions.

Third, the union's policy depends on the nature of the new technique. If the change results in a large and immediate reduction in the number of jobs, the union may elect to oppose it. However, the effect on the number of jobs in the bargaining unit tends to be more important than the effect on the number of jobs in the particular department affected by the change. Thus, although a particular change may reduce the number of jobs in one department, a local union covering the entire plant may not make an important issue of it. Also, the change in the degree and kind of skills required by the new process is of great importance, since it may transfer the work to a new craft or result in a conflict between two crafts over which one will do the work.[11]

If it becomes clear that some workers must be displaced, the union's reaction tends to occur in three stages. First, an attempt is made to maintain the employment and earnings of existing

[11] *Ibid.*

job holders. Second, once the union recognizes that a decrease in jobs and earnings cannot be prevented, it advocates transitional measures to reduce the shock of displacement. For example, attempts are made to widen the seniority unit to include interplant, interfirm, and interarea transfers as a matter of right for displaced employees; also, requests are made for advance notice of shutdowns. Third, the union claims that the job losses should be compensated for by a financial settlement in exchange for which management obtains greater freedom in deploying manpower.[12]

## 7. Make-Work Rules and Featherbedding

We have noted that, when the adoption of a new technique reduces the number of workers that are needed, unions sometimes resist the employer's efforts to dispense with the unneeded employees. The resulting make-work rules take various forms. Limits may be placed on the load that a worker may handle, and restrictions may be placed on the duties of workers. Some work may have to be done twice, the best-known example of this being the "bogus" rule of the International Typographical Union.[13] Modern equipment may be prohibited, and excessive crews may be required. For example, the railroad running crafts have sought—and in many states, obtained—full crew laws that regulate the size and composition of crews. The loss of output attributable to deliberately restrictive practices cannot be estimated with any accuracy. In some industries, like entertainment and railroad operation, there is a considerable amount of useless labor; in other industries, it is a minor problem. The reason why only a minority of unions pursue make-work policies to any considerable extent is that only a minority of their membership would benefit even temporarily from such policies.

The ability of unions to impose make-work rules depends

[12] J. Barbash, "The Impact of Technology on Labor-Management Relations," in G. Somers, E. Cushman, and N. Weinberg, *op. cit.*

[13] P. Weinstein, *Featherbedding and Technological Change*, Boston: D. C. Heath, 1965.

primarily on two things—the extent of the union's control of the market and the willingness of the firm to stand shutdowns. A union has control of the market if it is able to impose the rule on all competitors, thus putting none at a relative disadvantage. The willingness of a firm to take a shutdown depends on the costs involved, which themselves depend on whether the product is such that the business lost during the shutdown is largely a permanent loss that cannot be recovered later. If the product is of this type, as in entertainment, the water front, building construction, and transportation, there is a greater likelihood that make-work practices can be imposed. Although it is difficult to measure the importance of make-work rules at various points in time, there seems to be a feeling that they are less important now than in the period immediately after World War II. In part, this is due to the fact that such practices grew rapidly during 1937–1947. This period included several prewar years when firms, unfamiliar with the new unions, were learning to deal with them; the wartime years, when there was great pressure for uninterrupted production; and the first postwar years when there was a seller's market. After 1947, managements began to reform their rate structures and the classification of jobs. As one would expect, these changes did not occur without conflicts between management and labor.[14]

It should not be assumed, however, that make-work rules are entirely a product of unions. On the contrary, the roots of some of these rules are buried in the nineteenth century. Moreover, unorganized as well as organized workers engage in such practices. Several decades ago, Veblen claimed that the "conscientious withdrawal of efficiency" was common among all classes of society. He thought it unfortunate that the term "sabotage" was applied exclusively to the violent activities of organized workers, and did not include the "deliberate malingering, confusion, and misdirection of work" engaged in by employers and workers alike.[15]

[14] S. Slichter, J. Healy, and R. Livernash, *op. cit.* All other things equal, the pressure for make-work rules—and their importance—is likely to be less during periods of relatively full employment.

[15] T. Veblen, *The Engineer and the Price System,* 1921. Also see S. Matthewson, *Restriction of Output Among Unorganized Workers,* New York: Viking, 1931.

There seems to be little interest in the direct use of public policy to outlaw featherbedding. In the last few years, the Supreme Court has held that a union's attempt to maintain jobs in the face of employer opposition is a "legitimate labor dispute" and therefore cannot be restricted by injunction. In the Taft-Hartley Act, Congress made it an unfair labor practice for a labor organization "to cause . . . an employer to pay . . . for services which are not performed"; but this proved abortive. One of the most important problems is the absence of a legally workable definition of featherbedding. A number of states have passed tougher laws on this subject, but when tested they have invariably proved unconstitutional. The primary grounds for the negation of these laws has been that they interfere too much with the rights of unions to bargain collectively.[16]

## 8. Expenditures by Federal Departments and Agencies

The Federal government plays an extremely important role in the promotion of technological change and scientific advance in the United States. In the final three sections of this chapter, we discuss the magnitude, nature, and rationale of this role. We begin by looking at the amount spent by various Federal departments and agencies on research and development. What agencies account for most of the spending? What is the purpose of the R and D they support? How has the relative importance of various agencies shifted over time? Since the Federal government finances a large proportion of the nation's research and development, these are very important questions, which must be answered in some detail.

In 1966, almost half of all Federal R and D expenditures were made by the Department of Defense (Table 5.1), the primary purpose being to provide new and improved weapons and tech-

[16] See B. Aaron, "Governmental Restraints on Featherbedding," *Stanford Law Review,* 1953. Also see W. Gomberg, "The Work Rules and Work Practices Problem," *Labor Law Journal,* July 1961.

niques to promote the effectiveness of the armed forces. The largest expenditures were made by the Air Force; the smallest were made by the Army. Only about 25 percent was spent on research (rather than development), and this research was mainly in the physical and engineering sciences. The second and third largest spenders on R and D in 1966 were the National Aeronautics and Space Administration and the Atomic Energy Commission, both of which are also intimately connected with the cold war. Together with the Defense Department, they accounted for almost 90 percent of the R and D expenditures of the Federal government. About one quarter of the R and D carried out by NASA and AEC was research, most of it in the physical and engineering sciences.

In contrast with the Big Three, the fourth, fifth, and sixth largest spenders were not concerned primarily with national defense and the space race. The bulk of the R and D expenditures of the Department of Health, Education and Welfare (HEW)—the fourth largest spender—was related to the work of the National Institutes of Health. Consequently, most of the research expenditures of HEW were in the medical sciences. The fifth largest spender was the National Science Foundation (NSF), the general purposes of which are the encouragement and support of basic research and education in the sciences. Most of the NSF expenditures went for research in the physical and biological sciences. The sixth largest spender was the Department of Agriculture, where most of the R and D effort, which is coordinated with the research and educational activities of the land-grant colleges, was concerned with the production, utilization, and marketing of farm and forest products.

These six department and agencies accounted for practically all of the Federal government's R and D expenditures in 1966. Table 5.1 shows the extent of the changes in the volume and pattern of Federal R and D spending in the past several decades. It is interesting to note that the Department of Agriculture spent more on R and D than did the Department of Defense in 1940. It is also important to note that the total amount of R and D financed by the Federal government in 1966 was over 200 times what it was in 1940 and over four times what it was

*TABLE 5.1 Federal Expenditures for Research and Development and R and D Plant, by Agency, Fiscal Years 1940–1966*

| Department or Agency | 1940 | 1948 | 1956 | 1964 | 1966 [a] |
|---|---|---|---|---|---|
| | | | (millions of dollars) | | |
| Agriculture | 29.1 | 42.4 | 87.7 | 183.4 | 257.7 |
| Commerce | 3.3 | 8.2 | 20.4 | 84.5 | 93.0 |
| Defense | 26.4 | 592.2 | 2,639.0 | 7,517.0 | 6,880.7 |
| Army [b] | 3.8 | 116.4 | 702.4 | 1,413.6 | 1,452.1 |
| Navy [b] | 13.9 | 287.5 | 635.8 | 1,724.2 | 1,540.0 |
| Air Force [b] | 8.7 | 188.3 | 1,278.9 | 3,951.1 | 3,384.4 |
| Defense agencies | — | — | — | 406.9 | 464.5 |
| Department-wide funds | — | — | 21.9 | 21.1 | 39.7 |
| Health, Education and Welfare [c] | 2.8 | 22.8 | 86.2 | 793.4 | 963.9 |
| Interior | 7.9 | 31.4 | 35.7 | 102.0 | 138.7 |
| Atomic Energy Commission | — | 107.5 | 474.0 | 1,505.0 | 1,559.7 |
| Federal Aviation Agency | — | | — | 74.0 | 73.4 |
| National Aeronautics and Space Administration [d] | 2.2 | 37.5 | 71.1 | 4,171.0 | 5,100.0 |
| National Science Foundation | — | — | 15.4 | 189.8 | 258.7 |
| Office of Scientific Research and Development | — | 0.9 | — | — | — |
| Veterans Administration | — | — | 6.1 | 34.1 | 45.9 |
| All other agencies | 2.4 | 11.8 | 10.4 | 39.7 | 66.1 |
| Total | 74.1 | 854.7 | 3,446.0 | 14,693.9 | 15,437.7 |

Source: Federal Funds for Science XIV (*National Science Foundation, 1965*), *Table C-46. These figures are not entirely comparable with those in Chapter III. In this and other tables in this book, the totals may differ from those that are given because of rounding errors.*

[a] Estimates.

[b] Includes pay and allowances of military R and D personnel beginning in 1953 and support from procurement appropriations of development, test, and evaluation beginning in 1954.

[c] Federal Security Agency prior to 1952.

[d] National Advisory Committee on Aeronautics prior to 1958.

in 1956. Much of this increase has been due to wartime and postwar increases in spending on defense and space. The Defense Department's expenditures on R and D rose greatly during

World War II and continued to increase during the fifties.[17] In the early postwar period, the Atomic Energy Commission was established and its R and D expenditures grew rapidly. During the late fifties and early sixties, NASA's budget grew enormously. By the mid-sixties, Federal R and D spending had developed into the pattern shown in the last column of Table 5.1, there being a tremendous emphasis on defense and space technology.

## 9. The Rationale for the Major Federal Research and Development Programs

We saw in Chapter III that the Federal government supports most of the nation's research and development. Why is this necessary? Can we not rely on free enterprise to allocate resources efficiently in the area of R and D? Without government support, would not the right sorts of R and D be carried out? To begin with, one must consider those activities where it is technically impossible or grossly inefficient to deny the benefits of the activity to a citizen who is unwilling to pay the price; examples are national defense and the space program. In these "collective consumption" activities, the Federal government is the sole or principal purchaser of the equipment used to perform the function. Since it has the primary responsibility for these activities, it must also take primary responsibility for the promotion of technological advance in relevant areas. Few people would deny that the Federal government has this responsibility; indeed, the most rock-ribbed conservatives insist that, in matters of national defense and national prestige, the Federal government insure that a proper rate of technological change is maintained. Of course, because the Federal government has the principal responsibility in these areas, it does not necessarily follow that it must support most of the relevant research and develop-

[17] The leveling off of the Defense Department's R and D expenditures during the last few years covered in Table 5.1 was due mainly to the transition out of the development stage of the major ballistics missile programs such as the Atlas, Titan, and Minuteman.

ment. Heeding the example of some private firms, it might elect instead to encourage others to finance the R and D and to buy the fruits in the form of improved products. In fact, the Federal government follows this strategy in many areas of the public sector. But in the area of defense, such a strategy would be unthinkable because the risks would be so great, both to the nation and to firms supplying military equipment. Also, in the area of space exploration, this strategy has been rejected because of the size and riskiness of the space project.

Not all the major Federal R and D programs are designed to promote technological change in goods and services provided by the public sector of the economy. Some programs are designed to offset imperfections in the system that would otherwise lead to an under-investment in R and D. For example, the value of an improvement due to R and D may be less to the individual citizen than to the community as a whole. In part, this is the rationale for Federal support of health research: society gains over and above the gains of the individual citizen as health standards rise, the most obvious case being the reduction of contagious diseases. Other programs designed, at least in part, to offset imperfections in the system are those of the National Science Foundation, the nonmilitary parts of the Atomic Energy Commission, and the Department of Agriculture. To justify these programs, economists often assert that there are considerable discrepancies between the private and social benefits (and costs) of R and D in these areas. They maintain that these discrepancies occur because the results of R and D can be appropriated only to a limited extent and because, in some cases, of the riskiness and costliness of the research and development.[18]

The case for Federal support of basic research seems reasonably strong. Industrial firms almost certainly will invest less than is socially optimal in basic research. This is because the results of such research are unpredictable and usually of little direct value to the firm supporting the research, although potentially

[18] See K. Arrow, "Economic Welfare and the Allocation of Resources for Invention," *The Rate and Direction of Inventive Activity*, Princeton, N.J.: Princeton University Press, 1962; and R. Nelson, "The Simple Economics of Basic Scientific Research," *Journal of Political Economy*, June 1959.

of great value to society as a whole. With regard to agricultural research, Federal support could be defended originally on the grounds that the smallness of firms in farming would result in smaller R and D expenditures than would be socially desirable. As industries producing agricultural supplies and equipment grew, this argument was weakened. However, there remain certain aspects of farming that these companies do not touch. With regard to nonmilitary research on atomic energy, the expense of such research, as well as the dangers of monopoly, are often cited as reasons for Federal support. However, there has been considerable controversy over the role of the Federal government in the atomic energy programs.[19]

The choice of major research priorities is largely a political one. There are no simple cost-benefit criteria, for instance, which will decide whether health R and D expenditures should be increased at the expense of military R and D expenditures. The major choices are largely value judgments. Thus, the existing government R and D programs, like other government programs, are based only partly on purely economic criteria. Political pressures, the activities of special-interest groups, as well as historical accident, also play a role. It is difficult to see how this could be otherwise, since many variables that are of crucial economic importance cannot be measured very accurately. Ideally, one might like to base decisions on the marginal social rate of return from various types of research and development. Unfortunately, however, no one has found a way to measure these rates of return at all precisely.

## 10. Relations Between the Private and Public Sectors

Before World War II, the bulk of the research and development financed by the Federal government was done in government

[19] For example, see D. Price, *Government and Science,* New York: New York University, 1954, pp. 84–86; and D. Lilienthal's article in *International Science and Technology,* June 1963.

laboratories, relatively little being contracted out to industry or universities. For example, in the procurement of military aircraft, an open competition was held; the winning firm recovered its development costs in the form of profits on the sale of the airplanes, and the losing firms did not recoup their R and D investment. This situation changed radically during the war; and by the sixties, only about 20 percent of federally-financed R and D was carried out by government laboratories, the rest being contracted out to industry and universities. The agencies that rely most heavily on in-house laboratories are the Departments of Commerce, Agriculture, and Interior, all of which perform over two thirds of their R and D in such laboratories. In the Defense Department, the Army and Navy, with facilities like the Army Ballistics Research Laboratories, the Naval Ordnance Test Station, and the Fort Monmouth Signal Laboratory, depend more heavily on in-house laboratories than does the Air Force.[20]

The tremendous increase in Federal contracting has resulted in a blurring of the distinction between the private and public sectors. Some major contractors, particularly in the defense and space fields, do practically all of their business with the government. In many cases, their products—aircraft, missiles, and the like—have no civilian markets, much of their capital is provided by the government, and the government has agents involved in the managerial and operating structure of their organizations. Because of the great uncertainties involved in military R and D and the impossibility of competitive bidding for R and D contracts, the mechanism of the free market has been replaced largely by administrative procedure and negotiation. In a sense, as Don Price has pointed out, the government "has learned to socialize without assuming ownership." [21] Moreover, the need to maintain an industrial mobilization base, as well as political pressures, makes it likely that contracts will be awarded without strict attention to past performance.

[20] Almost two-thirds of the Atomic Energy Commission's R and D is carried out in Federal contract research centers administered by industry and non-profit institutions. These laboratories are not included as in-house in the NSF statistics.

[21] D. Price, *The Scientific Estate,* Cambridge, Mass.: Belknap Press, 1965, p. 43.

Some observers are troubled by the government's tendency to contract out the management of vital national scientific programs, such as the ballistic missile program. In 1954, the Air Force placed full responsibility for system engineering and technical direction in the hands of Space Technology Laboratories, a division of Ramo-Wooldridge Corporation. When Congressional criticism mounted, the Air Force transferred these functions to Aerospace Corporation, a nonprofit firm established to render advisory services to the government. In reviewing the program, the Comptroller General concluded, "By delegating the technical aspects of this program to a contractor, the Air Force has, to a significant degree, removed itself from the direct management of the program and, as a practical matter, has shifted a portion of its responsibility for the success of this crucial program to a contractor. We believe that a program of this importance should be conducted under the direct leadership and responsibility of the Government agency to which it is entrusted." [22]

Some people also question the propriety of the government's contracting out basic strategic studies. Nonprofit groups like the RAND Corporation, the Institute for Defense Analysis, and the Hudson Institute, have been created since World War II to provide research and advice for the armed services and the Secretary of Defense on questions of military strategy and tactics, as well as on related economic and political matters. In some respects, it may be unfortunate that such delicate questions have been placed in the hands of people who are only indirectly accountable to public scrutiny; but unless Congress is willing to increase the levels of compensation of Federal employees and to provide other inducements, it is difficult to see how the necessary expertise can be assembled in government agencies.

During the fifties, there was evidence of a trend toward reduced competence of government laboratories. Contractors often were able to provide better salaries, better facilities, and better administrative support, with the result that government laboratories had more difficulty in attracting and holding first-class

[22] Comptroller General of the United States, *Initial Report on Review of Administrative Management of the Ballistic Missile Program of the Air Force*, p. 2 of transmittal letter.

people. Moreover, it often seemed that contractors were given more significant and interesting assignments than the government laboratories. During the sixties, efforts were made to improve the work environment for scientists and technicians within the government. According to Adam Yarmolinsky, the "effects of the serious erosion of competence in our in-house laboratories which we faced in the Fifties have not yet been completely overcome. But the trend has been reversed." [23]

In 1962, the Bell Report,[24] produced by a top-level committee of Federal officials, was submitted to the President. This report stipulated that three criteria should be used to decide whether or not a particular R and D task should be contracted out. First, the top management and control of the Federal R and D effort must be firmly in the hands of full-time government officials.[25] Second, the basic rule should be to assign a job where it can be done most effectively and efficiently. Third, efforts should be made to avoid possible conflicts of interests, such as cases where a firm provides technological advice regarding a weapon system for which it later seeks a development or production contract. An important point made in the report was that, if the contract system is to work effectively, the government must be a sophisticated buyer. Efforts must be made by the government to maintain a staff of able and trained officials. The Bell Report received considerable attention and led to significant changes, but many observers believe that its full ramifications have yet to be felt.

During the late sixties, there was a trend toward the Federal

[23] A. Yarmolinsky, "Science Policy and National Defense," *American Economic Review,* May 1966. However, in recent years, there has been some indication that some government laboratories have been slow to adjust to changing needs. For example, see Research and Technical Programs Subcommittee, House of Representatives, *A Case Study of Utilization of Federal Laboratory Resources,* U.S. Government Printing Office, 1966, and *Science,* December 23, 1966.

[24] Bureau of the Budget, *Report to the President on Government Contracting for Research and Development,* April 30, 1962. Also see D. Price, *The Scientific Estate, op cit.,* pp. 38–39.

[25] To maintain this control, the government obviously needs enough in-house R and D capability to help it judge the value of proposals from the private sector.

government's financing a smaller percentage of the research and development performed by industry. From 1956 to 1967, the Federal government always financed over one-half of the research and development performed by industry, the percentage sometimes going up almost to 60. But in the late sixties, this percentage fell below 50, and in 1970 it was estimated to be 44.[26] This is a very substantial shift. Relative to a decade or so ago, industry is financing a considerably larger proportion of the R and D that it carries out.

[26] National Science Foundation, *National Patterns of R and D Resources*, Washington, 1969.

# CHAPTER SIX

# Public Policy and Technological Change

## 1. A Time of Inquiry and Appraisal

The past decade has witnessed a significant increase in the amount of attention devoted to public policies concerning technological change. In Congress, at least four subcommittees have launched investigations to determine the proper role of the Federal government in the support and management of research and development.[1] The President has established a commission which has studied the patent system and recommended ways in which it should be reformed. The National Academy of Sciences has

[1] There are a number of reasons for these investigations. Federal R and D expenditures, about 15 percent of the Federal administrative budget in the mid-sixties, are large enough to be politically visible. The rapid expansion of the Federal science budget seems to have come to an end, thus making it harder than in the past to "avoid" problems of choice. More than half of the nation's scientists and engineers are supported, directly or indirectly, by Federal R and D expenditures. In addition, of course, there is a growing recognition of the importance of science and technology, the public stake in them, and the increasing interdependence of the whole structure of science and technology.

been asked to evaluate the state of basic science in the United States and the techniques currently used by the government to select scientific programs. Many of the state governments have formed scientific advisory committees to help in the formulation of relevant state policies. In this chapter, we discuss some of the important policy issues that emerge from these and other investigations. To begin with, we look at various questions regarding the patent system, as well as relevant aspects of our antitrust policies. Then we discuss military development policies and Federal support for civilian technology.[2]

## 2. The Patent System

One of the major instruments of national policy regarding technology is the patent system. The United States patent laws grant the inventor exclusive control over the use of his invention for seventeen years, in exchange for his making the invention public knowledge. Not all new knowledge is patentable. A patentable invention "is not a revelation of something which existed and was unknown, but the creation of something which did not exist before." [3] "There can be no patent upon an abstract philosophical principle." [4] A patentable invention must have as its subject matter a physical result or a physical means of attaining some result, not a purely human means of attaining it. Moreover, it must contain a certain minimum degree of novelty. " 'Improvement' and 'invention' are not convertible terms . . . [W]here the most favorable construction that can be given . . . is that the article constitutes an improvement over prior inventions, but it embodies no new principle or mode of operation not utilized before by other inventors, there is no invention." [5]

[2] The discussion of public policy issues is not confined to this chapter. For example, some discussion of issues concerning labor displacement and adjustment to new techniques, as well as government R and D expenditures, is presented in Chapter V.

[3] Pyrene Mfg. Co. v. Boyce, C.C.A.N.J., 292 F. 480.

[4] Boyd v. Cherry, 50F. 279, 282.

[5] William Schwarzwaelder and Co. v. City of Detroit, 77F. 886, 891.

What proportion of patented inventions are used commercially, and how

Since Congress passed the original patent act in 1790, the arguments used to justify the existence of the patent laws have not changed very much. First, these laws are regarded as an important incentive to induce the inventor to put in the work required to produce an invention. Particularly in the case of the individual inventor, it is claimed that patent protection is a strong incentive. Second, patents are regarded as a necessary incentive to induce firms to carry out the further work and make the necessary investment in pilot plants and other items that are required to bring the invention to commercial use. If an invention became public property when made, why should a firm incur the costs and risks involved in experimenting with a new process or product? Another firm could watch, take no risks, and duplicate the process or product if it were successful. Third, it is argued that, because of the patent laws, inventions are disclosed earlier than otherwise, the consequence being that other inventions are facilitated by the earlier dissemination of the information. The resulting situation is often contrasted with the intense secrecy with regard to processes which characterized the medieval guilds and which undoubtedly retarded technological progress and economic growth.

Not all economists agree that the patent system is beneficial.

great are the private returns from them? The Patent, Coypright, and Trademark Foundation of George Washington University selected a 2 percent random sample of patents granted in 1938, 1948, and 1952, and determined (by interviews or correspondence with the inventor or assignee) the utilization status of the inventions sampled. According to the results, 51 percent of the patents assigned at date of issue to large companies were used commercially, 71 percent of those assigned at date of issue to small companies were used commercially, and 49 percent of the patents unassigned at date of issue were used commercially. For mechanical inventions, 57 percent were used commercially; whereas for electrical and chemical inventions, 44 percent were used commercially.

Not all of the patents in commercial use are profitable to their owners. The Patent, Copyright, and Trademark Foundation obtained information on profitability from assignees on 127 of 292 inventions reported in past or current use. Two thirds of those used in the past (for which there were data) were profitable; one third showed losses. Nine tenths of those currently in use (for which there were data) were profitable; one tenth showed losses. Dollar estimates of profits and losses were provided by assignees for 93 patents in current or past use. For the results, see B. Sanders, "Patterns of Commercial Exploitation of Patented Inventions by Large and Small Companies," *Patent, Copyright, and Trademark Journal,* Spring 1964.

A patent represents a monopoly right, although as many inventors can testify, it may represent a very weak one. Critics of the patent system stress the social costs arising from the monopoly. After a new process or product has been discovered, it costs little or nothing for other persons who could make use of this knowledge to acquire it. The patent gives the inventor the right to charge a price for the use of the information, the result being that the knowledge is used less widely than is socially optimal.[6] Critics also point out that patents have been used to create monopoly positions which were sustained by other means after the original patents had expired; they cite as example the aluminum, shoe machinery, and plate glass industries. In addition, the cross licensing of patents often has been used by firms as a vehicle for joint monopolistic exploitation of their market.

Critics also question the extent of the social gains arising from the system. They point out that the patent system was designed for the individual inventor, but that over the years most research and development has become institutionalized. They assert that patents are not really important as incentives to the large corporation, since it cannot afford to fall behind in the technological race, regardless of whether or not it receives a patent. They point out that, because of long lead times, most of the innovative profits from some types of innovations can be captured before imitators have a chance to enter the market. Moreover, they claim that firms keep secret what inventions they can, and patent those that they cannot.

Surveys of business firms provide mixed answers about the im-

[6] Since the use of knowledge by one individual does not reduce the ability of another individual to use it, it would be socially desirable in a static sense for existing knowledge to be available for use wherever it is of social value. However, if there were no financial reward for the knowledge producer, there would be less incentive to produce new knowledge. To promote optimal use while preserving this incentive, it has been suggested that society would be better off if the profits obtained by the patent holder could be awarded to him as a lump sum and if there were unrestricted use of the information. Unfortunately, this proposal suffers from important practical difficulties. See E. Mansfield, "Economics, Public Policy and the Patent System," *Journal of the Patent Office Society,* May 1965; and "National Science Policy," *American Economic Review,* May 1966, as well as section 3 below.

portance of patents in encouraging R and D and innovation. Some firms feel that patents are extremely important and that their research and innovative activities could not be sustained without them; others feel that patents make little difference in their search for, and introduction of, new products and processes. The electronics, chemicals, and drug industries make extensive use of patents; the automobile, paper, and rubber industries do not. Patents are much more important for independent inventors and small firms than for large firms, which are better able to carry out their inventive and innovative activities without the protection of the patent system. They are much more important for major product inventions than for process inventions (which can often be kept secret for a considerable period of time) or for minor product inventions.

Do the benefits derived from the patent system outweigh its costs? Like many broad issues of public policy, the facts are too incomplete and too contaminated by value judgments to permit a clear-cut, quantitative estimate of the effects of the patent system. Nonetheless, with or without such an estimate, it is impossible to avoid the relevant policy issue, and when confronted with it, there are few leading economists, if any, who favor abolition of the patent system. Even those who publish their agnosticism with respect to the system's effects admit that it would be irresponsible, on the basis of our present knowledge, to recommend abolishing it. Nonetheless, many economists seem to be interested in altering some of its features—which is hardly surprising in view of the fact that the patent system has continued for over a century without major change.

## 3. Proposed Changes in the Patent System

Over the years, there have been many proposals for reforming the U.S. patent system. Some have proposed that the length of the patent be varied in accord with the importance of the invention or the cost of making it. This proposal seems reasonable, since it is highly improbable that a single inflexible system can apply optimally both to major technological breakthroughs and to more routine, incremental changes. However, there are administrative difficulties arising from the problem of deciding the relative importance of inventions. Others have proposed that patent holders be required to pay an annual registration fee, the penalty for nonpayment being the premature termination of the patent. The purpose of this proposal, which has been put in effect in several countries (including the United Kingdom), is to weed out worthless patents.

Turning to more radical proposals, a system of general compulsory licensing has sometimes been suggested, which would permit everyone to obtain licenses under any patent. Under this system, patentees could no longer hope for attractive monopoly profits, but only for royalties from their licenses and cost advantages over their royalty-paying competitors. This proposal has been resisted almost everywhere because of the difficulties of determining "reasonable royalties" and because of a fear that inventive and innovative activity would be unduly discouraged.[7] Another proposal, put forth by Michael Polanyi, would "supplement licenses of right by government rewards to patentees on a level ample enough to give general satisfaction to inventors."[8] Every inventor would have the right to claim a public reward, and all other persons could use the invention freely. Because of the difficulty in setting individual awards, this proposal has received only a limited amount of attention.

It seems unlikely that fundamental changes of this sort will be made. If changes come about, they are likely to be less sweeping. For example, there has been some interest in a "delayed" examination" system, like that adopted in 1964 by the Dutch. Under the present system in the United States, the Patent Office

[7] An important facet of the controversy over compulsory licensing is the charge of suppression of patents which has been persistently repeated and angrily rejected.

[8] M. Polanyi, "Patent Reform," *Review of Economic Studies*, Summer, 1944. Proposals for systems of prizes and bonuses to inventors, rather than patents, are very old. For example, Alexander Hamilton suggested that the Federal government award prizes for important inventions. However, the problem of selecting inventors and inventions has limited the attention accorded this sort of scheme.

examines all patent applications in order of their filing date as to formalities, and sees whether the three essential statutory requirements of novelty, utility, and non-obviousness of the subject matter over the prior art are met.[9] Under a delayed examination system, an application is first examined only as to formal matters. If the application is in order, it is published as a "provisional" patent; and a full examination is made as to novelty and invention only if certain fees are paid by an interested party within a certain period of time. If full examination is not requested within this period, the patent will lapse.

The delayed examination system has some important advantages. Since the invention is published more quickly than under the present system, the technological information embodied in the patent will be disseminated more rapidly. This is socially desirable, though not always beneficial to the inventor. In addition, the provisional patent should be enough for patent applications filed chiefly for defensive purposes, that is, out of fear that if a patent is not obtained, someone else may obtain a patent on the same thing and use it against the original inventor. Thus, the delayed examination system should relieve the Patent Office of the burden of giving these applications a full examination, and this in turn should help to solve one of the Patent Office's oldest and most important problems—the long delay in issuing patents. In recent years, the average patent is issued after 3.5 years before the Patent Office.

There is also a question regarding the desirability of our interference procedure, which has long been criticized as overly time consuming and complex. The United States and Canada are the only countries where the question of priority among claimants of a single invention is adjudicated within the Patent Office. In most countries, the first applicant is granted the patent, and later applicants can go to court to have the question of priority determined. In addition, there is a question as to whether or not we in the United States are correct in allowing employee-inventors to assign away all their rights in patents re-

[9] By "formalities" is meant compliance with various rules of application. "Novelty," with some exceptions, means being the first to invent. "Invention" is a contribution above the exercise of ordinary skill.

lating to the business of the employer or resulting from work done for the employer, in consideration of employment. In West Germany, for example, the employed inventor receives under certain circumstances an extra compensation for his invention. Very little is known about the effects of such schemes on the performance of the employee-inventor or on the employer's incentives to carry out research and development. In the United States, there is little indication that existing practices will be altered in the forseeable future.

In 1966, the President's Commission on the Patent System recommended a number of changes in the legal superstructure of the system, the purpose being to speed patent approval, tighten the standards of patentability, and reduce the cost of challenging and defending patent rights. It recommended that, when two or more persons separately apply for a patent on the same invention, the patent should issue to the one who was first to file his application. Moreover, applicants should be responsible for knowing all prior inventions and developments anywhere in the world. One consequence of these proposals, if adopted, would be the abolition of interference proceedings. In addition, the commission recommended that the existence of supposedly new technology should be publicized by automatically printing patent applications within two years from the filing date. An inventor or firm could keep the invention secret only by withdrawing an application or not filing one in the first place.

The Commission also recommended that standby statutory authority be provided for an optional delayed examination system whereby the examination is deferred at the option of the applicant. Any outside party would be allowed to challenge any patent application by paying a fee for Patent Office experts to look into claims to prior development. Also, losers in a patent challenge would be required to pay the entire cost of the litigation, treble damages would be allowed in some infringement suits, and a district court ruling on the validity of a patent would apply everywhere in the country. (The last recommendation prevents a patent owner who loses a claim in one court from harassing an opponent by filing suit in another jurisdiction. At present, this can be done.) The period of patent protection would be changed to twenty years from date of first application, and computer languages and programs would not be patentable. Legislation will be required before these changes can be made.[10]

## 4. Antitrust Policy

The antitrust laws are designed to promote competition and to control monopoly. For example, the Sherman Act of 1890, the first major federal legislation directed against monopoly, outlaws conspiracies or combinations in restraint of trade and forbids the monopolizing of trade or commerce. The courts have had the difficult job of deciding what business conditions are actually forbidden by these acts. At the time of the U.S. Steel decision in 1920, the Supreme Court interpreted monopolization to mean market conduct which tends to coerce rivals, and held that "the law does not make mere size an offense or the existence of unexerted power an offense." This interpretation stood for two decades, but was altered in the Alcoa case of 1945, when the Court said that control by a firm of a large proportion of the market could by itself constitute a violation of the Sherman Act.

There are at least two fairly distinct approaches to antitrust policy. The first approach is concerned primarily with market performance—the industry's rate of technological change, efficiency, and profits, the conduct of individual firms, and so on. Advocates of this approach argue that, in deciding antitrust cases, one should review in detail the performance of the firms in question to see how well they have served the economy. This test, as it is usually advocated, relies heavily on evaluation of the technological "progressiveness" and "dynamism" of the firms in question. Unfortunately, it is difficult to know how such an

[10] *To Promote the Progress of the Useful Arts in an Age of Exploding Technology,* Report of the President's Commission on the Patent System, Washington, D.C., 1966. For a discussion of some of the adverse reaction to the proposed changes, see "Criticism of Patents Bill Is Expected," *New York Times,* April 16, 1967.

evaluation is to be carried out, since there is no way of telling at present whether the rate of technological change in a particular industry represents "good" or "bad" performance. In view of the vagueness of the criteria and the practical realities of the antitrust environment, the adoption of this test would probably be an invitation to nonenforcement.

The second approach emphasizes the importance of an industry's market structure—the number and size distribution of buyers and sellers in the market, the ease with which new firms can enter, and the extent of product differentiation. According to this approach, one should look to market structure for evidence of undesirable monopolistic characteristics. Although many economists favor this approach, others claim that a vigorous antitrust policy based on this test would be a mistake; for example, as pointed out in previous chapters, Schumpeter and Galbraith assert that large firm size and a high level of concentration are conducive to rapid technological change and rapid utilization of new techniques. If true, this is an extremely important point. But is it true? Does the evidence indicate that an industry dominated by a few giant firms is generally more progressive than one composed of a larger number of smaller firms? Some of the relevant findings have been presented in previous chapters. Although the evidence is extremely limited, it should be brought together at this point and examined.

## 5. Firm Size, Market Structure, Technological Change, and the Utilization of New Techniques

To prevent confusion, it is advisable to distinguish among several related issues. First, what is the effect of an industry's market structure on the amount it spends on R and D? Suppose that, for a market of given size, we could replace the largest firms by a larger number of somewhat smaller firms—and thus decrease concentration. If the largest firms in this industry were giants, like Standard Oil of New Jersey or U.S. Steel, the available evidence, which is extremely tentative, does not suggest that total R and D expenditures are likely to decrease considerably. On the contrary, there is usually no tendency for the ratio of R and D expenditures to sales to be higher among the giants than among their somewhat smaller competitors. However, if the largest firms in the industry were considerably smaller than this or if concentration were reduced greatly, one might expect a decrease in R and D expenditures, because firm size often must exceed a certain minimum for R and D to be profitable.[11]

Second, to what extent would such a change in market structure be harmful because of economies of scale in R and D? Obviously, the answer to this question varies with the type of research or development being considered. In research, the optimal size of group may be fairly small in many areas; for example, the transistor, the maser, the laser, and radio command guidance were conceived by an individual or groups of not more than a dozen persons. In development, the optimal size of effort tends to be larger, particularly in the aircraft and missile industries where tremendous sums are spent on individual projects.[12] However, in most industries, the limited data that are available do not seem to indicate that only the largest firms can support effective R and D programs; there is generally no indication that the largest programs have any marked advantage over somewhat smaller ones.[13]

Third, is there any evidence that R and D programs of given scale are carried out more productively in large firms than in small ones? The data are extremely limited, but they seem to

[11] See Chapter III, and F. Scherer, Testimony Before Senate Subcommittee on Antitrust and Monopoly, May 18 and 25, 1965.

[12] Development costs are sometimes very high in other industries too. For example, DuPont spent $25 million to develop Corfam, and $50 million for Delrin. (However, these figures may be increased somewhat by the cost of pilot plants, some of whose output is sold commercially.) See D. Stillerman's and F. Scherer's testimony before the Senate Subcommittee on Antitrust and Monopoly, May 18, 1965.

[13] See E. Mansfield, *Industrial Research and Technological Innovation*, New York: W. W. Norton & Company, Inc., 1968, Chapter II; and J. Jewkes, D. Sawers, and R. Stillerman, *The Sources of Invention*, New York: St. Martin's Press, Inc., 1959.

indicate that, in most industries for which we have information, the answer is no. In most of these industries, when the size of R and D expenditures is held constant, increases in size of firm are associated with decreases in inventive output. The reasons for this are by no means obvious. Some observers claim it is because the average capabilities of technical people are higher in the smaller firms, R and D people in smaller firms are more cost conscious than those in larger firms, and the problems of communication and coordination tend to be less acute in smaller firms.[14]

Fourth, what is the effect of an industry's market structure on how rapidly new processes and products, both those developed by the industry and those developed by others, are introduced commercially? The answer seems to depend heavily on the types of innovations that happen to occur. If they require very large amounts of capital, it appears that the substitution of fewer large firms for more smaller ones may lead to more rapid introduction; if they require small amounts of capital, this may not be the case. Another important factor is the ease with which new firms can enter the industry. If increased concentration results in increased barriers to entry, it may also result in slower application of new techniques, since innovations often are made by new firms. Indeed, many new firms are started for the specific purpose of carrying out innovations[15]

Fifth, what is the effect of an industry's market structure on how rapidly innovations, once they are introduced, spread through an industry? The fact that large firms tend to be quicker than small firms to introduce a new technique does not imply that increased concentration results in a faster rate of diffusion. On the contrary, the very small amount of evidence in Chapter IV bearing on this question seems to suggest that greater concentration in an industry may be associated with a slower rate of

[14] E. Mansfield, *ibid.*, and A. Cooper, "R and D Is More Efficient in Small Companies," *Harvard Business Review*, June 1964.

[15] See E. Mansfield, *ibid.*, Chapter VI; R. Schlaifer's testimony before the Senate Subcommittee on Antitrust and Monopoly, May 25, 1965; and G. Brown, "Characteristics of New Enterprises," *New England Business Review*, June–July, 1957.

diffusion. However, the observed relationship is weak and could well be due to chance[16]

Contrary to the allegations of Galbraith, Schumpeter, and others, there is little evidence that industrial giants are needed in all or even most industries to insure rapid technological change and rapid utilization of new techniques. Moreover, there is no statistically significant relationship between the extent of concentration in an industry and the industry's rate of technological change, as measured by the methods described in Chapter II.[17] Of course, this does not mean that industries composed only of small firms would necessarily be optimal for the promotion and diffusion of new techniques. On the contrary, there seem to be considerable advantages in a diversity of firm sizes, no single firm size being optimal in this respect. Complementarities and interdependencies exist among large and small firms. There is often a division of labor, smaller firms focusing on areas requiring sophistication and flexibility and catering to specialized needs, bigger firms concentrating on areas requiring large production, marketing, or technological resources. However, there is little evidence in most industries that firms that are considerably smaller than the biggest firms are not big enough for these purposes.

## 6. Antitrust Policy and the Patent System

The patent system, if left unchecked by a vigorous antitrust policy, can be made an effective device for the spread of monopoly power. In 1790, when the first patent law was enacted, invention was a matter of individual tinkering, and the inventor's limited talents, energy, and financial resources made it very unlikely that he could monopolize an industry. However, during the next century, these safeguards were substantially weakened.

[16] E. Mansfield, *ibid.*, Chapter VII.

[17] E. Mansfield, *ibid.*, *Chapter* IV; and N. Terleckyj, *Sources of Productivity Advance*, Ph.D. Thesis, Columbia University, 1960.

By the first decade of this century, "the patent system had become a special sanctuary for trusts, pools, and trade confederacies," in the judgment of two leading students of the problem.[18]

Over the past thirty years, there has been a discernible trend toward resolving conflicts between the patent system and antitrust policy in favor of the latter. The mortality rate for patents before the courts has increased substantially, and the mortality rate has varied directly with the rank of the court. From 1948 to 1954, the district courts ruled on 664 patents in reported cases and found 355 invalid and 108 not infringed; during the same period, the Supreme Court threw out five of the seven patents that came before it. Moreover, Congress has also set higher standards for patentability. In the early statutes, the test was whether the invention was "new and useful." In the Patent Act of 1952, a patent may not be obtained "if the subject matter as a whole would have been obvious at the time the invention was made to a person having ordinary skill in the art to which said subject matter pertains."

In addition, the courts have curtailed the extension of the effects of the patent beyond the invention described in the patent claim. In the forties, the Supreme Court established the doctrine that the license could not fix the prices of unpatented products produced by patented processes. Also, it ruled against a patent provision based on the sale of an unpatented as well as a patented commodity. Unless they result in industry price fixing, price restrictions established by the patentee on the licensee are generally held to be legal, but in recent years there has been considerable effort to reverse this doctrine. The courts have also shown a tendency to deny the patentee exclusive use in cases where it results in substantial monopoly power. Between 1941 and 1957, over 100 judgments provided for compulsory licensing or outright dedication. The rationale for these decisions is illustrated by the statement of Judge Forman in the General Electric case: "In view of the fact that General Electric achieved its dominant position in the industry and maintained it in great

[18] G. Stocking and M. Watkins, *Monopoly and Free Enterprise*, New York: The Twentieth Century Fund, 1951, p. 454.

measure by its extension of patent control the requirement that it contribute its existing patents to the public is only a justified dilution of that control made necessary in the interest of free competition in the industry."[19]

## 7. The Acquisition of New Weapons

Since the Department of Defense accounts for almost half of the Federal R and D budget, its policies regarding the utilization of R and D resources are obviously of great importance. As noted in Chapter III, one of the most important characteristics of the weapons acquisition process is its uncertainty. A weapons system as it is initially conceived is generally quite different from the weapons system which actually emerges from development. For example, a study of six fighter plane development projects shows that four of the six ended up with engines that differed from the original plans, three had different electronic systems, five had to be modified extensively and three came out of development essentially different airplanes. Changes of this sort occur because engineers try to squeeze the last ounce of performance out of their systems, because the experts' consensus can be quite wide of the mark, because unanticipated technical problems arise, and sometimes because of poor planning. Besides these uncertainties which originate largely in the technological character of weapons development, there are other uncertainties regarding the demand for a particular weapons system. Because of unexpected changes in the rate of progress of related technologies and unexpected changes in the nature of opposing forces, a weapons system may be much less valuable than originally estimated. The

[19] 115 F. Supp. 835 (D.N.J. 1953). This section relies heavily on J. Markham, "Inventive Activity: Government Controls and the Legal Environment," in *The Rate and Direction of Inventive Activity,* National Bureau of Economic Research, 1962. Note that the tax laws can also influence the rate of technological change. For example, research and development was encouraged by the 1954 changes in the tax laws which permitted R and D expenditures to be deducted as a current expense rather than being treated as a capital investment.

importance of these uncertainties is magnified, of course, by the vast size and extended duration of major weapons programs.

Because of these uncertainties, the market system, in anything like its customary form, has not been applied to the acquisition of new weaponry. Instead, the government has exercised control over sellers through the auditing of costs and through the intimate involvement of its agents in the managerial and operating structure of the sellers. Moreover, there is extensive government ownership of the sellers' facilities, the government decides what weapons are to be created through its program decisions, and payments to the sellers are frequently based on costs incurred (although, as we shall see, simple cost-plus pricing has become much less important than it used to be). Clearly, "competition" in this environment does not mean what it does in the market system. Another important difference between the weapons acquisition process and the development process in most other areas of the economy is the extent of the technological advance that is attempted. The attempt in military development for rapid and major advances on a broad front have little counterpart in the normal activities of commercial industries.

Several other characteristics of the nation's weapons contractors should also be noted. First, the market position of a particular contractor is rather insecure. Product lines change rapidly to meet changing requirements of new technology, and it has been relatively easy for newcomers to join the ranks. One problem that has sometimes occurred is that firms have directed their efforts in such a way as to enhance their capability for new projects at the expense of good work on current projects. Second, the critical resources of a weapons contractor are scientific and engineering talent rather than the more conventional inputs noted in Chapter II. Third, performance on past programs has not always been given sufficient weight in the selection of contractors. In part, of course, this is because opinions may vary on the quality of past performance and because good performance in the past does not necessarily mean good performance in the future. In addition, however, source selection involves multiple objectives and other economic and political considerations may play an important role.

How well have our weapons programs been carried out? Studies made in the early sixties suggest that, although technical performance, reliability, and development time have been at least reasonably satisfactory, there has been a notable failure to hold development and production costs to reasonable levels. In part, this has been due to inadequate attention being given to the efficient use of technical and other manpower and to the development of increments of technical performance and other features that are not worth their cost ("gold-plating"). More fundamentally, it has been due to the greater emphasis that the services have placed on time and quality considerations than on cost reduction, the result being that contractors recognized that a record of meeting schedules with good products was much more important in getting new business than a reputation for low costs. Moreover, cost-plus contracts provided little incentive to reduce costs. During the sixties, efforts were made by Secretary of Defense McNamara to emphasize cost reduction to a much greater extent than was formerly the case.

## 8. Military Development Policy

Based on a long series of studies of military development projects, economists at the RAND Corporation, led by Burton Klein, have made a number of suggestions regarding military development policy. In their view,[20] it is important that the government devote a very significant proportion of its military R and D expenditures to activities falling outside the major weapons systems programs, that is, to basic research, exploratory development, and advanced development. By developing in this way a large menu of technology, we can hope to buy at a relatively low price the capability to adapt our weapons programs to the actual strategic situation in a short period of time. According to Glennan,[21] there is a particular need for more ad-

[20] B. Klein, *op. cit.*

[21] T. Glennan, "Research and Development," in S. Enke, *Defense Management,* Englewood Cliffs, N.J.: Prentice-Hall, Inc., 1967.

vanced development, which is the first point in the evolution of a system where military needs are confronted with available and potential technology. The prototype hardware of various kinds that are constructed in advanced development are possible building blocks in the development of operational systems. Apparently, an underinvestment of this sort occurs because of the nature of existing procedures and the preferences of program managers.

In carrying out weapons systems programs, Klein [22] suggests that a frankly experimental approach be adopted. Requirements for systems should be stated initially in broad terms, flexibility should be maintained, and decisions on the best set of compromises should be postponed until there is a reasonable basis for making them. Components should be tested as soon as possible, and the integration of the system should be postponed until the major uncertainties have been reduced substantially. Since he believes that the uncertainties involved are very great initially but that they diminish substantially as a project proceeds, he argues that the optimal strategy to overcome difficult technological problems often is to run in parallel several approaches designed to serve the same end. This is in contrast to the type of development strategy which emphasizes the integration, to the maximum extent possible, of the total process of development and production, the entire process being viewed as a single planning problem to be dealt with as a whole from the beginning. In the past, the latter strategy has been important in the acquisition of new weapons.

One of the important conclusions of the RAND studies is that, in planning military development, there has been a tendency to underestimate and suppress uncertainty. For example, the services have tended to specify their requirements for advanced weapons systems too early, too optimistically, and in too great detail. Also, there has been evidence in some programs of a commitment to production tooling at too early a stage of the development. In addition, problems of other kinds have been

[22] Klein, *op. cit.* For some criticism of Klein's ideas, see M. Peck and F. Scherer, *The Weapons Acquisition Process*, Boston: Harvard University Press, 1962.

cited, some stemming from the way in which contractors are selected and rewarded. In contrast to earlier days when there was competition among prototypes, competition now occurs at the design stage, because the development of more than one model is considered too expensive. Unfortunately, there are great uncertainties at the design stage and companies have a natural tendency to be optimistic, the result being that it is difficult to make a wise selection. Moreover, once the decision has been made, there is no more competition (although there may be considerable rivalry among alternative systems). The services are locked to a sole source, which ordinarily carries out production as well as development.

Robert McNamara, when Secretary of Defense during the sixties, attempted to improve the situation through the institution of a number of changes in procedures and organization, two of the most important innovations being the program definition phase and incentive contracts. During the program definition phase, later known as contract definition, competing contractors pursue alternative paths toward defining the development effort to be carried out and identifying the design specifications for the end product. At the end of this phase, they submit bids containing cost estimates for development and procurement. The point of the program definition phase was to postpone commitment to procurement of a specific item until the major uncertainties that can be resolved have been resolved. Moreover, the program definition phase was meant to permit the rival claims of competing contractors to be tested more rigorously than by an examination of their sales presentations.

Changes have also been made in the types of contractual arangements used in military research and development. Until the sixties, these contracts were typically cost-plus-fixed-fee (CPFF), this kind of contract being defended on the ground that the "product" was so unpredictable that the risks to the seller would require a very high fixed price if it were feasible at all. Recognizing that there is little incentive to reduce costs in CPFF contracts, the Department of Defense began to switch to incentive contracts during the early sixties. Since they reward cost reduction by giving the firm a certain percentage of the difference between its actual costs and the negotiated target costs, incentive contracts are likely to lead to greater efficiency than CPFF contracts, if the target costs are the same. However, many observers challenge the assumption that the target costs are the same, pointing out that the contractor, with considerable advantages in the negotiation of target costs, has more incentive to increase these costs under incentive contracts.[23]

A more drastic proposal has been made by Carl Kaysen,[24] who argues that, to a greater extent, military research *and* development should be divorced from production, and performed in nonprofit research institutes and government laboratories, which in his view are better suited than business firms to carry out this task. If development could be separated from production, it would be possible to reap the benefits from competition at the development stage and from a freer choice of suppliers and contract instruments at the production stage. A fundamental consideration in judging this proposal is the cost involved in separating development from production. It is sometimes argued that these costs are high, because there is considerable overlap and similarity between these two functions and because learning is transferred between them. We need much better estimates of these costs, as well as the possible benefits. In its present form, this scheme may be so at odds with political reality as to be of limited practical significance, but it raises a number of interesting questions.

Finally, there has been considerable concern over the concentration of military R and D expenditures in a relatively few firms. In 1964, three firms—North American, General Dynamics, and Lockheed—received 23 percent of the Defense Department's R and D money; and ten firms received 53 percent. This tendency is not confined to the DOD; for example, in 1963, three firms received 37 percent of NASA's R and D money, and ten firms received 61 percent. Critics assert that this concentration of R and D funds promotes undue concentration of production

[23] For a discussion of the post-McNamara management of the Department of Defense, see *Science,* February 13, 1970.

[24] C. Kaysen, "Improving the Efficiency of Military Research and Development," *Public Policy,* 1963.

and employment because the firm that receives an R and D contract generally receives the follow-on production contract and because the research will sometimes be of benefit to the commercial work of the performing firm. Also, and this is a somewhat different point, some observers feel that military R and D should be split up into a larger number of smaller pieces, particularly at the research phase, where there is less likelihood of important economies of scale.

## 9. Federal Support for Research and Development in Transportation, Housing, and Pollution Control

Turning from the military and space efforts, we find a widespread feeling that as a nation the United States is underinvesting in certain types of research and development. For example, Nelson had stated that "aside from the fields of defense and space, peacetime atomic energy, and perhaps public health, it is likely that we are relying too much on private incentives as stimulated by the market to generate R and D relevant to the public sector . . . [Also,] aside from the fields of defense and space, there probably is too little research and experimentation aimed at exploring radically new techniques and ways of meeting needs . . . Surely we can do better than to rely so heavily on 'spillover' from defense and space to open up the really new possibilities in materials, energy sources, etc."[25]

Three areas often cited as needing more research and development are transportation, housing, and pollution of air and water. With regard to transportation, our cities suffer from congestion, commuting to work is often difficult and time-consuming, the accident toll is considerable, and delays in terminals are high. According to the critics, new transportation technologies point toward the solution of many of these problems, but their po-

[25] R. Nelson, "Technological Advance, Economic Growth, and Public Policy," *RAND Corporation* P-2835, December 1963, p. 17.

tential benefits are not being realized, because of unresolved organizational, administrative, and financial problems; because of the failure to take a more integrated look at transportation as a whole; and because the resources devoted to far-reaching R and D in this area are meager.

With regard to housing, there is a feeling in many quarters that the industry is backward technologically and that more advanced technologies should be explored in an effort to reduce housing costs. The impediments to the development and use of new techniques are numerous, the typical construction firm being too small to carry out its own R and D, and the industry being fragmented into various types of trades and subcontractors. Moreover, outmoded building codes bar many types of innovation, the codes being protected by various special interest groups and the fragmented character of local governments.

There is considerable public concern regarding increases in air and water pollution. The growth of urban populations has concentrated the discharge of wastes into a small sector of the atmosphere, and resulted in increased contamination. A similar pollution of water resources has taken place. "We have been unbelievably irresponsible in contaminating our water resources to the point where we are now faced with a problem of limited supply. . . . As our population density has increased, the natural cleansing ability of streams has been exceeded." [26] To combat air and water pollution, the President's Science Advisory Committee and the National Commission on Automation have recommended that an enlarged research program be carried out.

Two things should be noted regarding the alleged deficiency in R and D expenditures in these areas. First, one cannot make any estimate of the adequacy or inadequacy of R and D expenditures in a given field by looking simply at society's evaluation of the importance of the activity. In addition, one must consider the probability and cost of achieving a significant improvement in the activity through research and development. No matter how important a particular goal may be, if more research and development are unlikely to help us achieve it,

[26] National Commission on Technology, Automation, and Economic Progress. *Technology and the American Economy,* Washington, D.C., 1966.

there is no reason to increase our R and D expenditures in this area. Second, there has been a feeling in some quarters that a lack of promising, well-developed research ideas and of receptivity to change in these areas is responsible for the low level of R and D spending. According to President Johnson's Science Advisor, "what we lack in many of the civilian problem areas . . . is not a consensus on their importance. Rather, it is a lack of solid R and D program proposals. . . . We cannot buy and create progress in a field which is not ready to progress." [27] Also, there has been a feeling that the need is for use of techniques already available, rather than for more R and D. For example, Capron "would place federally supported R and D fairly low on the list of things we need in the fields of urban housing and urban transportation. . . . Our problems in these areas are much more institutional and organizational." [28]

From President Nixon's budget request in 1970, it appears that some change is occurring in priorities. More money will be allocated for research on environmental and social problems rather than military and space programs. Indeed, research and development expenditures by the Department of Defense and NASA will be lower in 1971 than in 1970. Also, there seems to be some tendency for more to be spent on applied research and less to be spent on basic research.[29]

## 10. Federal Support for Civilian Technology

Transportation, housing, and pollution control are not the only areas considered to suffer from an underinvestment in research and development. According to the Council of Economic Advisers, "in a number of industries the amount of organized private research undertaken is insignificant, and the technology

[27] Testimony of D. Hornig in *The Research and Development Programs: The Decision Making Process,* Hearings before a Subcommittee of the House Committee on Government Operations, January 1966, p. 6.

[28] W. Capron, *ibid.,* p. 19.

[29] See *Business Week,* February 14, 1970, pp. 90–92.

of many of these low research industries has notably failed to keep pace with advances elsewhere in the economy." [30] Freeman, Poignant, and Svennilson conclude that, "in spite of the great increase in research and development activity, there are good reasons for believing that in many cases this activity is still below the level desirable for efficient and sustained economic growth." [31]

In 1963, the Department of Commerce proposed a Civilian Industrial Technology program to encourage and support additional R and D in industries that it regarded as lagging. It proposed that support be given to important industries, from the point of view of employment, foreign trade, and so forth, which have "limited or dispersed technological resources." Examples cited by the department included textiles, building and construction, machine tools and metal fabrication, lumber, foundries and castings. The proposal met with little success on Capital Hill. Industrial groups opposed the bill because they feared that government sponsorship of industrial R and D could upset existing competitive relationships.[32] More recently, Nelson, Peck, and Kalachek [33] have suggested that a National Institute of Technology be established to provide grants for research and development aimed at placing the technology of various industries on a stronger scientific footing and to test the feasibility and attributes of advanced designs. In their view, work of this sort, which falls between basic research and product development, is likely to be in need of additional support. In cases where a broad-scale systems view is required but is deterred by the smallness of existing firms and the fragmentation of market interest, the institute would also support work through the middle and later stages of development.

Unfortunately, there is little evidence to support or deny the

[30] Council of Economic Advisers, *Annual Report*, 1964, p. 105.

[31] C. Freeman, M. Poignant, and S. Svennilson, *Science, Economic Growth, and Government Policy*, Organization for Economic Cooperation and Development, 1963, p. 42.

[32] See U.S. Department of Commerce, *The Civilian Industrial Technology Program*, Washington, D.C., 1963; and D. Allison, *op. cit.*

[33] R. Nelson, M. Peck, and E. Kalachek, *Technology, Economic Growth, and Public Policy*, Washington, D.C.: The Brookings Institution, 1966.

belief that the areas in question suffer from an underinvestment in R and D. Since we cannot estimate the social returns from additional R and D of various sorts at all accurately, it is difficult to make a strong case one way or the other. The proponents of additional government support for civilian technology rely heavily on the argument that R and D generates significant external economies and that, under these circumstances, private initiative is unlikely to support work to the extent that is socially optimal. However, this argument only suggests that the government or some other organization not motivated by profit should support some R and D in these areas; it does not tell us whether such support is currently too large or too small.

Under these circumstances, perhaps the most sensible strategy, both in connection with the proposed Institute of Technology and some of the other programs discussed above, is to view the relevant policy issues in the context of the theory of sequential decision making under uncertainty. To the extent possible, R & D programs should be begun on a small scale and organized so as to provide data regarding the returns from a larger program. On the basis of the data that result, a more informed judgment can be made regarding the desirability of increased—or in some cases, decreased—programs of Federal support. A strategy of this sort has been suggested by Nelson, Peck, and Kalachek, as well as the present author. If this approach is adopted, it is important that proper attention be given to devising methods by which the results of the small-scale program are to be measured. Without such measures, the sequential approach will obviously be of little use.

## 11. The Levelling-Off of Government R and D Expenditures, Changing Priorities, and Technological Assessment

Beginning in the late sixties, the public's attitude toward science and technology seemed to change, enthusiasm waning and wariness growing. In part, this has been due to anti-war sentiments, many observers, following the example of President Eisenhower in his last address to the nation, fearing undue power and influence by a "military-industrial complex" allied with—and having considerable control over—the nation's scientists and technologists. To an increasing degree, science and technology have been pictured as the handmaiden of the cold warriors. Moreover, to make matters worse, the firms and agencies carrying out defense R and D have been accused of inefficiency and waste by Senator Proxmire and others, the cost-overruns on the Lockheed C-5A and the deficiencies of the TFX having made the headlines.

In addition, more and more emphasis has been placed on the costs associated with technological change in the civilian economy. For example, it is frequently pointed out that the application of modern technology has resulted in air and water pollution. Technological advances that were once heralded as modern miracles are now viewed with suspicion, and sometimes banned, as their unexpected side effects cause problems and damage and even death. For example, insecticides turn out to kill birds and fish as well as harmful insects; and synthetic sweeteners may result in cancer.

This growing skepticism toward science and technology has coincided with the tightening of Federal fiscal constraints caused by the Indochina war and the accompanying inflation, the result being that Federal expenditures on research and development have remained essentially constant at about $16 billion per year from 1967 to 1971.[34] After the long period of spectacular increases (described in Chapters III and V), this leveling-off has forced painful readjustments in many firms, government agencies, and universities. It has contributed to a marked slowdown during the late sixties and early seventies in the rate of increase of total R and D expenditures in the United States.[35] Moreover, coupled with the increased supply of engineers and scientists

[34] National Science Foundation, *Science Resources Studies Highlights,* August 14, 1970.

[35] According to the National Science Foundation, total expenditures on research and development in the United States increased from about $24 billion in 1967 to about $27 billion in 1970. See *National Patterns of R and D Resources,* National Science Foundation, 1969.

being graduated by the nation's universities, it has helped to bring about an apparent surplus of some kinds of scientists and engineers.

In response to these and other developments, important changes have taken place in the allocation of Federal R and D expenditures, and public policy measures have been proposed to anticipate and curb the harmful side effects of technological change. With regard to the changes in the allocation of Federal R and D expenditures, there has been a tendency in recent years for nondefense, nonspace activities to get a greater share of Federal R and D expenditures. For example, defense and space accounted for 84 percent of total Federal R and D expenditures in 1965 but only 79 percent in 1970.[36] This is in response to, and evidence of, the recent change in social priorities and public opinion toward domestic problems and away from military work.

With regard to the design of public policy measures to anticipate and curb the harmful side effects of technological change, the topic of "technological assessment" has become fashionable in Washington. For example, Congressman Daddario has introduced a bill "to provide a method for identifying, assessing, publicizing, and dealing with the implications and effects of applied research and technology"[37] by establishing a Technology Assessment Board. The desirability of a mechanism for technology assessment is clear enough. But given our very limited capacity for technological forecasting and the tremendous problems of evaluating the various effects of a new technology, it is important that we proceed cautiously. We must guard against such a mechanism's turning into a device for the retardation of needed technological change.

The skepticism of the late sixties and early seventies is probably a healthy reaction to some of the naïvete and self-serving optimism of the earlier years. But it would be a tragic mistake to carry this reaction too far. Modern technology is a tremendously powerful force that can allow man to live a fuller, longer, hap-

[36] *Ibid.*

[37] *Technology Assessment,* House Subcommittee on Science and Astronautics, August 1968.

pier life. If we use this force with care, we can better our own lot as well as that of our less fortunate neighbors. Modern technology is no panacea, but used with wisdom it can accomplish a great deal.

# SELECTED READINGS

Abramowitz, M., "Resource and Output Trends in the United States since 1870," *American Economic Review,* May 1956.

Arrow, K., "The Economic Implications of Learning by Doing," *Review of Economic Studies,* June 1962.

———, "Economic Welfare and the Allocation of Resources for Invention," *The Rate and Direction of Inventive Activity,* Princeton, 1962.

Arrow, K., H. Chenery, B. Minhas, and R. Solow, "Capital-Labor Substitution and Economic Efficiency," *Review of Economics and Statistics,* August 1961.

Bright, J., *Research, Development, and Technological Innovation,* Richard D. Irwin, 1964.

Brown, M., *On the Theory and Measurement of Technological Change,* Cambridge University, 1966.

Brozen, Y., "The Future of Industrial Research," *Journal of Business,* October 1961.

Carter, A., "The Economics of Technological Change," *Scientific American,* April 1966.

Cherington, P., "Kaysen on Military Research and Development," *Public Policy,* 1963.

Comanor, W., "Research and Competitive Product Differentiation in the Pharmaceutical Industry in the United States," *Economica,* November 1964.

Denison, E., *The Sources of Economic Growth in the United States,* Committee for Economic Development, 1962.

Enke, S., *Defense Management,* Prentice-Hall, 1967.

Galbraith, J. K., *American Capitalism,* Houghton Mifflin, 1952.

———, *The New Industrial State,* Houghton Mifflin, 1967.

Griliches, Z., "Hybrid Corn: An Exploration in the Economics of Technological Change," *Econometrica,* October 1957.

Hamberg, D., "Invention in the Industrial Laboratory," *Journal of Political Economy,* April 1963.

Harrod, R., *Towards a Dynamic Economics,* London, 1948.

Hicks, J., *The Theory of Wages,* London, 1932.

Hitch, C., and R. McKean, *The Economics of Defense in the Nuclear Age,* Harvard, 1960.

Jantsch, E., *Technological Forecasting in Perspective,* Organization for Economic Cooperation and Development, 1966.

Jewkes, J., D. Sawers, and R. Stillerman, *The Sources of Invention,* W. W. Norton, 1970.

Jorgensen, D. and Z. Griliches, "The Explanation of Productivity Change," *Review of Economic Studies,* July 1967.

Kaysen, C., "Improving the Efficiency of Military Research and Development," *Public Policy,* 1963.

Kendrick, J., *Productivity Trends in the United States,* Princeton, 1961.

Klein, B., "The Decision Making Problem in Development," *The Rate and Direction of Inventive Activity,* Princeton, 1962.

———, "Policy Issues in the Conduct of Military Development Programs," *The Economics of Research and Development,* edited by R. Tybout, Ohio State, 1965.

Leontief, W., "On Assignment of Patent Rights on Inventions Made Under Government Research Contracts," *Harvard Law Review,* January 1964.

Machlup, F., *An Economic Review of the Patent System,* Study 15 of the Senate Subcommittee on Patents, Trademarks, and Copyrights, 1958.

Mansfield, E., *Industrial Research and Technological Innovation,* W. W. Norton for the Cowles Foundation for Research in Economics at Yale University, 1968.

———, *The Economics of Technological Change,* W. W. Norton, 1968.

———, *Numerical Control: Diffusion and Impact in the Tool and Die Industry,* Small Business Administration, 1971.

———, *Defense, Science, and Public Policy,* W. W. Norton, 1968.

———, J. Rapoport, J. Schnee, S. Wagner, and M. Hamburger, *Research and Innovation in the Modern Corporation,* W. W. Norton, 1971.

Markham, J., "Inventive Activity: Government Controls and the Legal Environment," *The Rate and Direction of Inventive Activity,* Princeton, 1962.

Marschak, T., T. Glennan, and R. Summers, *Strategy for R and D,* Springer-Verlag, 1967.

National Academy of Science, *Basic Research and National Goals,* House Committee on Science and Astronautics, 1965.

———, *Applied Science and Technological Progress,* House Committee on Science and Astronautics, 1967.

National Bureau of Economic Research, *The Theory and Empirical Analysis of Production,* New York, 1967.

National Commission on Technology, Automation, and Economic Progress, *Report to the President of the Commission,* 7 volumes, February 1966.

National Science Foundation, *National Patterns of R and D Resources, 1953–70,* Washington, 1969.

Nelson, R., "Uncertainty, Learning, and the Economics of Parallel Research and Development Efforts," *Review of Economics and Statistics,* 1961.

———, "The Link Between Science and Invention: The Case of the Transistor," *The Rate and Direction of Inventive Activity,* Princeton, 1962.

———, M. Peck, and E. Kalachek, *Technology, Economic Growth, and Public Policy,* Brookings, 1967.

Organization for Economic Cooperation and Development, *The Requirements of Automated Jobs,* Paris, 1965.

Peck, M. and F. Scherer, *The Weapons Acquisition Process,* Harvard, 1962.

Pessemier, E., *New-Product Decisions,* McGraw-Hill, 1966.

Price, D., *Government and Science,* New York University, 1954.

———, *The Scientific Estate,* Belknap Press, 1965.

Robinson, J., "The Classification of Inventions," *Review of Economic Studies,* 1937–8.

———, *The Accumulation of Capital,* Richard D. Irwin, 1956.

Rogers, E., *Diffusion of Innovation,* The Free Press of Glencoe, 1962.

Salter, W., *Productivity and Technical Change,* Cambridge University, 1960.

Scherer, F., "Government Research and Development Programs," *Measuring Benefits of Government Expenditures,* edited by R. Dorfman, Brookings, 1965.

———, "Firm Size, Market Structure, Opportunity, and the Output of Patented Inventions," *American Economic Review,* December 1965.

Schmookler, J., *Invention and Economic Growth,* Harvard, 1966.

Schon, D., *Technology and Change,* Delacorte, 1967.

Schultz, G. and A. Weber, *Strategies for the Displaced Worker,* Harper and Row, 1966.

Schumpeter, J., *Capitalism, Socialism, and Democracy,* Harper and Row, 1947.

———, *Business Cycles,* McGraw-Hill, 1939.

Slichter, S., J. Healy, and R. Livernash, *The Impact of Collective Bargaining on Management,* Brookings, 1960.

Solow, R., "Technical Change and the Aggregate Production Function," *Review of Economics and Statistics,* 1957.

———, "Investment and Technical Change," *Mathematical Models in the Social Sciences,* ed. by Arrow, Karlin, and Suppes, Stanford, 1959.

———, *The Nature and Sources of Unemployment in the United States,* Almquist and Wiksell, 1964.

Somers, G., E. Cushman, and N. Weinberg, *Adjusting to Technological Change,* Harper and Row, 1963.

Terleckyj, N., *Sources of Productivity Advance,* Ph.D. Thesis, Columbia University, 1960.

United States Bureau of the Budget, *Report to the President on Government Contracting for Research and Development,* April 30, 1962.

United States Department of Commerce, *Technological Innovation,* Washington, January 1967.

Usher, A., *A History of Mechanical Inventions,* Harvard, 1954.

Veblen, T., *The Instinct of Workmanship in the State of the Industrial Arts,* Macmillan, 1914.

Vernon, R., "International Investment and International Trade in the Product Cycle," *Quarterly Journal of Economics,* May 1966.

Villard, H., "Competition, Oligopoly, and Research," *Journal of Political Economy,* December 1958.

Weinstein, P., *Featherbedding and Technological Change,* D. C. Heath, 1965.

# INDEX